The Third Kind:

A Compendium of U.F.O. Encounters

By

Michael Ryan

This book is dedicated to my family, friends and everyone who has supported my work… even though they did not always understand what it was about...

This book contains references to Recorded UFO events throughout history as well as unexplained phenomenon which by description where or are considered to be UFO encounters. In the end, you decide…

Preface

This book is a compendium of all recorded UFO sightings and encounters I could collect from both modern and ancient times.

All Material here in is a collection of recorded events of UFO and unexplained anomalies that are believed to be UFO activity, all information in the book is collected from various archives and locations but has been verified as actual sightings or encounters from witnesses throughout history from 214 BC to 2015AD.

This book will also contain some implied conclusions drawn up from connections made from events that in some way seem to interact, but in the end all information is speculative and is there for up to you to decide.

Enjoy.

Contents

Michael Ryan

Early UFO Sightings

As early as 214 BC ancient romans have reported flying ships in the sky. Titus Livius Patavinus or Livy (in English) records a number of portents in the winter of this year, including *navium speciem de caelo adfulsisse* ("an appearance of ships had shone forth from the sky").

This is The First known Example of U.F.O.'s recorded in history, recorded by a roman historian, while creating a complete history of Rome and roman culture and life style.

In 74 BC flame-like "pithoi" from the sky was reported, According to Plutarch, a Roman army commanded by Lucullus was about to begin a battle with Mithridates VI of Pontus when "all on a sudden, the sky burst asunder, and a huge, flame-like body was seen to fall between the two armies. In shape, it was most like a wine-jar, and in color, like molten silver." Plutarch reports the shape of the object as like a wine-jar (pithos). The apparently silvery object was reported by both armies.

70 BC A bright light over the Temple in Jerusalem. On a later day, "ὤφθη μετέωρα περὶ πᾶσαν τὴν χώραν ἅρματα καὶ φάλαγγες ἔνοπλοι διᾴττουσαι τῶν νεφῶν καὶ κυκλούμεναι τὰς πόλεις" ("there appeared in the air over the whole country chariots and armed troops coursing through the clouds, surrounding the cities").

Note: It is interesting to note that UFO's have been seen over the temple throughout history since then, and one very recently in 2011.

150 AD 100 foot "beast" accompanied by a "maiden". On a sunny day near the Via Campana, a road connecting Rome and Capua, a single witness, probably Hermas the brother of Pope Pius I, saw "a 'beast' like a piece of pottery (ceramos) about 100 feet in size, multicolored on top and shooting out fiery rays, landed in a dust cloud, accompanied by a "maiden" clad in white. Vision 4.1-3. in The Shepherd of Hermas.

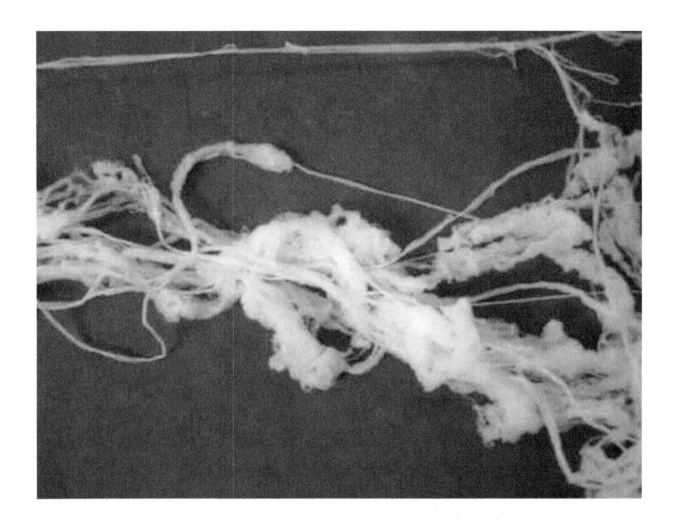

196 AD angel hair. Historian Cassius Dio described "A fine rain resembling silver descended from a clear sky upon the Forum of Augustus." He used some of the material to plate some of his bronze coins, but by the fourth day afterwards the silvery coating was gone.

Michael Ryan

1500's to 1600's

04-14-1561 celestial phenomenon over Nuremberg. At sunrise on the 14th April 1561, the citizens of Nuremberg beheld "A very frightful spectacle." The sky appeared to fill with cylindrical objects from which red, black, orange and blue white disks and globes emerged. Crosses and tubes resembling cannon barrels also appeared whereupon the objects promptly "began to fight one another." This event is depicted in a 16th-century woodcut by Hans Glaser.

09-26-1609 shiny object like "a bowl/washbasin" making "a thunderous sound" and flying "fast like an arrow". Belatedly recorded in the Annals over a month late, on September 26, 1609, over "clear and cloudless" skies (three places recorded Sa hour (9-11 AM), one recorded Oh hour (noon), and one recorded Mi hour (1-3 PM)), a shiny object resembling "a bowl" or "a washbasin", suddenly appeared over the skies, made "a thunderous sound" and flew "fast like an arrow", and that "heaven and earth shook". It looked "as if it would land", but then it "tilted and rose", and then "it disappeared into sparks", with a comment that "it looked as if it was in the air by some energy".

1800's sightings

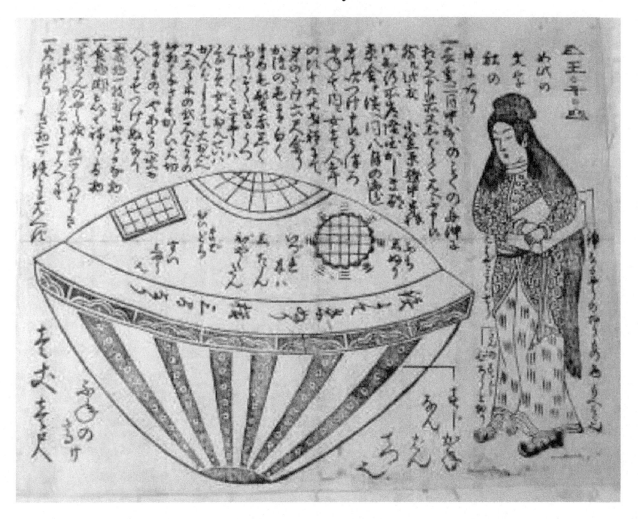

02-22-1803 Utsuro-bune at Haratono-hama. On February 22 (or March 24) in 1803 local fishermen reportedly saw a vessel drifting in close-by waters. They say when they investigated it, "a beautiful young woman" they described as having red and white hair and dressed in strange clothes appeared. The fisherman claim she held a square box "that no one was allowed to touch" and she spoke to them in a language they never heard before. Modern UFO believers think this story was a credible document of a close encounter of third kind in early Japan. Historians and Ethnologists consider it to be folklore.

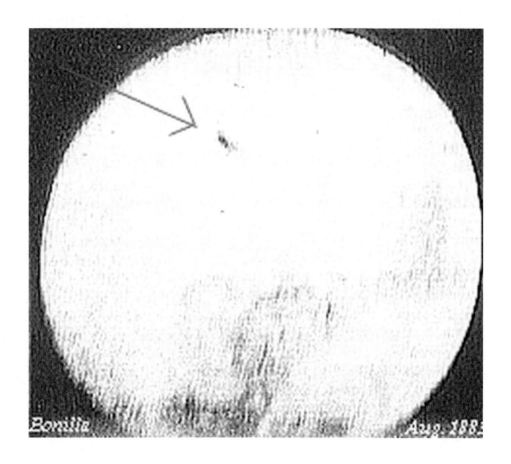

08-12-1883 José Bonilla Observation. On August 12, 1883, the astronomer José Bonilla reported that he saw more than 300 dark, unidentified objects crossing the sun disk while observing sunspot activity at Zacatecas Observatory in Mexico. He was able to take several photographs, exposing wet plates at 1/100 second. It was subsequently determined that the objects were highflying geese.

10-24-1886 Maracaibo. A letter from the US consulate in Maracaibo Venezuela was printed in the December 18, 1886 issue of *Scientific American* reporting a meteorological occurrence described as a bright light accompanied by a humming noise that caused occupants of a hut to become ill.

1896–1897 Mystery airships. Numerous reports of UFO sightings, attempted abductions that took place around the United States in a 2-year period.

04-17-1897 Aurora, Texas, UFO incident. A tale of a UFO crash and a burial of its alien pilot in the local cemetery was sent to newspapers in Dallas and Fort Worth in April 1897 by local correspondent S.E. Hayden

Discovery of Evidence Surrounding 1897 UFO Crash Baffles Scientists

During the 1896–1897 timeframe, numerous sightings of a cigar-shaped mystery airship were reported across the United States.

One of these accounts appeared in the April 19, 1897, edition of the *Dallas Morning News*. Written by Aurora resident S.E. Haydon, the alleged UFO is said to have hit a windmill on the property of a Judge J.S. Proctor two days earlier at around 6am local (Central) time, resulting in its crash. The pilot (who was reported to be "not of this world", and a "Martian" according to a reported Army officer from nearby Fort Worth) did not survive the crash, and was buried "with Christian rites" at the nearby Aurora Cemetery.

(The cemetery contains a Texas Historical Commission marker mentioning the incident.)

Reportedly, wreckage from the crash site was dumped into a nearby well located under the damaged windmill, while some ended up with the alien in the grave. Adding to the mystery was the story of Mr. Brawley Oates, who purchased Judge Proctor's property around 1935. Oates cleaned out the debris from the well in order to use it as a water source, but later developed an extremely severe case of arthritis, which he claimed to be the result of contaminated water from the wreckage dumped into the well. As a result, Oates sealed up the well with a concrete slab and placed an outbuilding atop the slab. (According to writing on the slab, this was done in 1957.)

In 1998, Dallas-based TV station KDFW aired a lengthy report about the Aurora incident. Reporter Richard Ray interviewed former *Fort Worth Star Telegram* reporter Jim Marrs and other locals, who said something crashed in Aurora. However, Ray's report was unable to find conclusive evidence of extraterrestrial life or technology. Ray reported that the State of Texas erected a historical plaque in town that outlines the tale and labels it "legend."

On December 2, 2005, *UFO Files* first aired an episode related to this incident, titled "Texas' Roswell". The episode featured a 1973 investigation led by Bill Case, an aviation writer for the *Dallas Times Herald* and the Texas state director of Mutual UFO Network (MUFON).

MUFON uncovered two new eyewitnesses to the crash. Mary Evans, who was 15 at the time, told of how her parents went to the crash site (they forbade her from going) and the discovery of the alien body.

Charlie Stephens, who was age 10, told how he saw the airship trailing smoke as it headed north toward Aurora. He wanted to see what happened, but his father made him finish his chores; later, he told how his father went to town the next day and saw wreckage from the crash.

MUFON then investigated the Aurora Cemetery, and uncovered a grave marker that appeared to show a flying saucer of some sort, as well as readings from its metal detector. MUFON asked for permission to exhume the site, but the cemetery association declined permission. After the MUFON investigation, the marker mysteriously disappeared from the cemetery and a three-inch pipe was placed into the ground; MUFON's metal detector no longer picked up metal readings from the grave, thus it was presumed that the metal was removed from the grave.

MUFON's report eventually stated that the evidence was inconclusive, but did not rule out the possibility of a hoax. The episode featured an interview with Mayor Brammer who discussed the town's tragic history.

20th Century Sightings

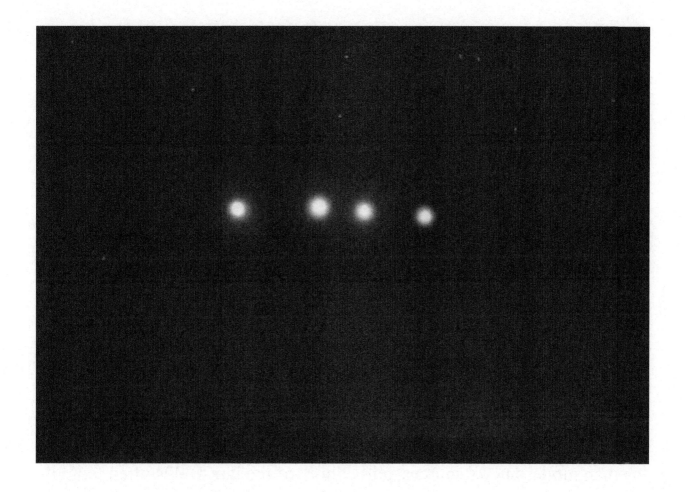

1909 **Mystery airships** or **phantom airships** are a class of unidentified flying objects best known from a series of newspaper reports originating in the western United States and spreading east during 1896 and 1897. According to researcher Jerome Clark, airship reports were made world-wide from the 1880s to 1890s. Mystery airship reports are seen as a cultural predecessor to modern claims of extraterrestrial-piloted flying saucer-style UFOs. Typical airship reports involved unidentified lights, but more detailed accounts reported ships comparable to a dirigible.

Reports of the alleged crewmen and pilots usually described them as human looking, although sometimes the crew claimed to be from Mars. It was popularly believed that the mystery airships were the product of some genius inventor not ready to make knowledge of his creation public. For example, Thomas Edison was so widely speculated to be the mind behind the alleged airships that in 1897 he "was forced to issue a strongly worded statement" denying his responsibility.

Author Gregory L. Reece has argued that mystery airships are unlikely to represent test flights of real human-manufactured dirigibles as no record of successful airship flights are known from the period and "it would have been impossible, not to mention irrational, to keep such a thing secret." To the contrary, however, there were in fact several functional airships manufactured before the 1896–97 reports (e.g., Solomon Andrews made successful test flights of his "Aereon" in 1863), but their capabilities were far more limited than the mystery airships. Reece and others note that contemporary American newspapers of the "Yellow journalism" era were more likely to print manufactured stories and hoaxes than modern news sources, and editors of the late 1800s often would have expected the reader to understand that such stories were phony. Period journalists did not seem to take airship reports very seriously, as after the major 1896–97 flap concluded the subject was not given further investigation and quickly fell from public consciousness. The airship reports received further attention only in the mid-twentieth century when UFO investigators suggested the airships might represent earlier precursors to post-World War II UFOs.

The Third Kind

Most people remember that in the middle of 1909 there were numerous reports from various parts of Otago, and a few from other parts of New Zealand, that people had seen strange lights in the air in the evening. Descriptions of the lights varied greatly, and theories to account for them were equally divergent. Some people said aurora; some said airships, and of the latter, a few even heard the aviators muttering darkly in "German." There were people who said they saw the supposed objects by daylight, when they were not luminous, but were apparently dark material masses. Those who did not see the phenomena usually tapped their foreheads, and said, "Poor chap"—the words being a gentle reflection upon the last-named observer of the mystery. One of these "poor chaps," who did not seem in the least afraid of the title, and disarmed his hearers by poking fun at himself, read a most interesting paper before the astronomical section of the Philosophical Society last evening. This was Mr. J. Orchiston, who, having himself seen the lights very clearly, not only recorded his own observation, but collected a number of descriptions by other people. The features of the lights, according to Mr. Orchiston, were their great brilliancy, high speed, low elevation above the ground (a few hundred feet), and the fact that they followed approximately the slope of the ground, rising over hills and dipping into valleys. His own estimate of their speed, made from reported observations at various places, was 200 miles per hour, and the light was strong enough for the time to be read on a watch when the luminous cloud was many miles away.

The paper naturally aroused a lively discussion. Mr. F. W. Furkert said that Mr. Orchiston deserved credit for his courage in facing such an audience with such a paper. If one wanted to be branded as a "40 horse-power liar" in these days, one only has to say "aerial lights." But he had seen one himself in the Catlins district, which was a dry area. Mr. Furkert described the light as appearing over a distant range of hills, and travelling at such a speed that its reappearance in a gap in the hills was correctly guessed at.

The members of the section discussed the matter from various points of view, and it was suggested by Professor Marsden that an observation by means of a spectroscope would have shown at once whether the lights were of auroral origin.

1917, 08-13, 09-13, 10-13 The **Miracle of the Sun** was an event which occurred just after midday on Sunday 13 October 1917, attended by some 30,000 to 100,000 people who were gathered near Fátima, Portugal. Several newspaper reporters were in attendance and they took testimony from many people who claimed to have witnessed extraordinary solar activity. This recorded testimony was later added to by an Italian Catholic priest and researcher in the 1940s.

According to these reports, the event lasted approximately ten minutes. The three children (Lúcia dos Santos, Jacinta Marto and Francisco Marto) who originally claimed to have seen Our Lady of Fátima also reported seeing a panorama of visions, including those of Jesus,

Our Lady of Sorrows, Our Lady of Mt. Carmel, and of Saint Joseph blessing the people.

The event was officially accepted as a miracle by the Roman Catholic Church on 13 October 1930. On 13 October 1951, the papal legate, Cardinal Tedeschini, told the million people gathered at Fátima that on 30 October, 31 October, 1 November, and 8 November 1950, Pope Pius XII himself witnessed the miracle of the sun from the Vatican gardens.

The people had gathered because three young shepherd children had predicted that at high noon the lady who had appeared to them several times would perform a great miracle in a field near Fátima called Cova da Iria. According to many witnesses, after a period of rain, the dark clouds broke and the sun appeared as an opaque, spinning disc in the sky. It was said to be significantly duller than normal, and to cast multicolored lights across the landscape, the people, and the surrounding clouds. The sun was then reported to have careened towards the earth in a zigzag pattern, frightening those who thought it a sign of the end of the world. Witnesses reported that their previously wet clothes became "suddenly and completely dry, as well as the wet and muddy ground that had been previously soaked because of the rain that had been falling".

Estimates of the number of people present range from between 30,000 to 40,000 by Avelino de Almeida, writing for the Portuguese newspaper *O Século*, to 100,000, estimated by Dr. Joseph Garrett, professor of natural sciences at the University of Coimbra, both of whom were present on that day. The event was attributed by believers to Our Lady of Fátima, a reported apparition of the Blessed Virgin Mary to the children who had made predictions of the event on 13 July 1917, 19 August, and 13 September. The children stated that the Lady had promised them that she would on 13 October reveal her identity to them and provide a miracle "so that all may believe."

The term **foo fighter** was used by Allied aircraft pilots in World War II to describe various UFOs or mysterious aerial phenomena seen in the skies over both the European and Pacific theaters of operations.

Though "foo fighter" initially described a type of UFO reported and named by the U.S. 415th Night Fighter Squadron, the term was also commonly used to mean any UFO sighting from that period. Formally reported from November 1944 onwards, witnesses often assumed that the foo fighters were secret weapons employed by the enemy.

The Robertson Panel explored possible explanations, for instance that they were electrostatic phenomena similar to St. Elmo's fire, electromagnetic phenomena, or simply reflections of light from ice crystals.

The first sightings occurred in November 1944, when pilots flying over Germany by night reported seeing fast-moving round glowing objects following their aircraft. The objects were variously described as fiery, and glowing red, white, or orange. Some pilots described them as resembling Christmas tree lights and reported that they seemed to toy with the aircraft, making wild turns before simply vanishing. Pilots and aircrew reported that the objects flew formation with their aircraft and behaved as if under intelligent control, but never displayed hostile behavior. However, they could not be outmaneuvered or shot down. The phenomenon was so widespread that the lights earned a name – in the European Theater of Operations they were often called "kraut fireballs" but for the most part called "foo-fighters". The military took the sightings seriously, suspecting that the mysterious sightings might be secret German weapons, but further investigation revealed that German and Japanese pilots had reported similar sightings.

On 13 December 1944, the Supreme Headquarters Allied Expeditionary Force in Paris issued a press release, which was featured in the *New York Times* the next day, officially describing the phenomenon as a "new German weapon". Follow-up stories, using the term "Foo Fighters", appeared in the *New York Herald Tribune* and the British *Daily Telegraph*.

In its 15 January 1945 edition *Time* magazine carried a story entitled "Foo-Fighter", in which it reported that the "balls of fire" had been following USAAF night fighters for over a month, and that the pilots had named it the "foo-fighter". According to *Time*, descriptions of the phenomena varied, but the pilots agreed that the mysterious lights followed their aircraft closely at high speed.

Some scientists at the time rationalized the sightings as an illusion probably caused by afterimages of dazzle caused by flak bursts, while others suggested St. Elmo's Fire as an explanation.

The "balls of fire" phenomenon reported from the Pacific Theater of Operations differed somewhat from the foo fighters reported from Europe; the "ball of fire" resembled a large burning sphere which "just hung in the sky", though it was reported to sometimes follow aircraft. On one occasion, the gunner of a B-29 aircraft managed to hit one with gunfire, causing it to break up into several large pieces which fell on buildings below and set them on fire. There was speculation that the phenomena could be related to the Japanese fire balloons' campaign. As with the European foo fighters, no aircraft was reported as having been attacked by a "ball of fire"

The postwar Robertson Panel cited foo fighter reports, noting that their behavior did not appear to be threatening, and mentioned possible explanations, for instance that they were electrostatic phenomena similar to St. Elmo's fire, electromagnetic phenomena, or simply reflections of light from ice crystals. The Panel's report suggested that "If the term "flying saucers" had been popular in 1943–1945, these objects would have been so labeled."

1941 One of the most mysterious stories of a crashed UFO with alien bodies preceded the well known Roswell event by some six years. This case was first brought to investigators by Leo Stringfield in his book "UFO Crash / Retrievals: The Inner Sanctum."He opened a tantalizing account of a military controlled UFO crash retrieval which is still being researched today. The details of the case were sent to him in a letter from one Charlette Mann, who related her minister-grandfather's deathbed confession of being summoned to pray over alien crash victims outside of Cape Girardeau, Missouri in the spring of 1941.Reverend William Huffman had been an evangelist for many years, but had taken the resident minister reigns of the Red Star Baptist Church in early 1941.

Church records corroborate his employment there during the period in question.

After receiving this call to duty, he was immediately driven the 10-15 mile journey to some woods outside of town. Upon arriving at the scene of the crash, he saw policemen, fire department personnel, FBI agents, and photographers already mulling through the wreckage.He was soon asked to pray over three dead bodies. As he began to take in the activity around the area, his curiosity was first struck by the sight of the craft itself.

Expecting a small plane of some type, he was shocked to see that the craft was disc-shaped, and upon looking inside he saw hieroglyphic-like symbols, indecipherable to him.

He then was shown the three victims, not human as expected, but small alien bodies with large eyes, hardly a mouth or ears, and hairless.

Immediately after performing his duties, he was sworn to secrecy by military personnel who had taken charge of the crash area. He witnessed these warnings being given to others at the scene also.

As he arrived back at his home at 1530 Main Street, he was still in a state of mild shock, and could not keep his story from his wife Floy, and his sons. This late night family discussion would spawn the story that Charlette Mann would hear from her grandmother in 1984, as she lay dying of cancer at Charlette's home while undergoing radiation therapy.

Charlette was told the story over the span of several days, and although Charlette had heard bits and pieces of this story before, she now demanded the full details.

As her grandmother tolerated her last few days on this Earth, Charlette knew it was now or never to find out everything she could before this intriguing story was lost with the death of her grandmother.

She also learned that one of the members of her grandfather's congregation, thought to be Garland D. Fronabarger, had given him a photograph taken on the night of the crash. This picture was of one of the dead aliens being help up by two men.

There are also Fire Department records of the date of the crash. This information does confirm the military swearing department members to secrecy, and also the removal of all evidence from the scene by military personnel.

Guy Huffman, Charlette's father also told the story of the crash, and had in his possession the photograph of the dead alien. He showed the picture to a photographer friend of his, Walter Wayne Fisk.

He has been contacted by Stanton Friedman, but would not release any pertinent information.

Charlette had no luck in getting Fish to return calls or answer letters. It has been rumored that Fisk was an advisor to the President, and if this

was the case, would account for his silence on the facts of the Missouri crash.

This case ends like many others, but appears by all indications to be authentic. All who have come in contact with Charlette Mann found her to be a trustworthy person who is not given to sensationalism, and has sought no gain from her account

There is still research being done on the Missouri crash, and hopefully more information will be forthcoming to validate this remarkable case.

1942 Hopeh Incident. This UFO was snapped by an American photographer in Tiensten, Hopeh province, China, in 1942. Several people in the photograph appear to be pointing up at the object.

THURSDAY MORNING.   Los Angeles Times   FEBRUARY 26, 1942.   **B**

Searchlights and Anti-aircraft Guns Comb Sky During Alarm

02-24-1942 The **Battle of Los Angeles**, also known as **The Great Los Angeles Air Raid**, is the name given by contemporary sources to the rumored enemy attack and subsequent anti-aircraft artillery barrage which took place from late 24 February to early 25 February 1942 over Los Angeles, California. The incident occurred less than three months after the United States entered World War II as a result of the Japanese Imperial Navy's attack on Pearl Harbor, and one day after the bombardment of Ellwood on 23 February.

Initially, the target of the aerial barrage was thought to be an attacking force from Japan, but speaking at a press conference shortly afterward, Secretary of the Navy Frank Knox called the incident a "false alarm." Newspapers of the time published a number of reports and speculations of a cover-up. Some modern-day UFOlogists have suggested the targets were extraterrestrial spacecraft. When documenting the incident in 1983, the U.S. Office of Air Force History attributed the event to a case of "war nerves" likely triggered by a lost weather balloon and exacerbated by stray flares and shell bursts from adjoining batteries.

Air raid sirens sounded throughout Los Angeles County on the night of 24–25 February 1942. A total blackout was ordered and thousands of air raid wardens were summoned to their positions. At 3:16 am the 37th Coast Artillery Brigade began firing .50 caliber machine guns and 12.8-pound anti-aircraft shells into the air at reported aircraft; over 1,400 shells would eventually be fired. Pilots of the 4th Interceptor Command were alerted but their aircraft remained grounded. The artillery fire continued sporadically until 4:14 am. The "all clear" was sounded and the blackout order lifted at 7:21 am.

Several buildings and vehicles were damaged by shell fragments, and five civilians died as an indirect result of the anti-aircraft fire: three killed in car accidents in the ensuing chaos and two of heart attacks attributed to the stress of the hour-long action. The incident was front-page news along the U.S. Pacific coast, and earned some mass media coverage throughout the nation.

Within hours of the end of the air raid, Secretary of the Navy Frank Knox held a press conference, saying the entire incident was a false alarm due to anxiety and "war nerves." Knox's comments were followed by statements from the Army the next day that reflected General George C. Marshall's belief that the incident might have been caused by commercial airplanes used as a psychological warfare campaign to generate panic.

Some contemporary press outlets suspected a cover up. An editorial in the *Long Beach Independent* wrote, "There is a mysterious reticence about the whole affair and it appears that some form of censorship is trying to halt discussion on the matter." Speculation was rampant as to invading airplanes and their bases. Theories included a secret base in northern Mexico as well as Japanese submarines stationed offshore with the capability of carrying planes. Others speculated that the incident was either staged or exaggerated to give coastal defense industries an excuse to move further inland.

Representative Leland Ford of Santa Monica called for a Congressional investigation, saying, "...none of the explanations so far offered removed the episode from the category of 'complete mystification' ... this was either a practice raid, or a raid to throw a scare into 2,000,000 people, or a mistaken identity raid, or a raid to lay a political foundation to take away Southern California's war industries."

A photo published in the Los Angeles Times on February 26, 1942 has been cited by modern-day conspiracy theorists and UFOlogists as evidence of an extraterrestrial visitation. They assert that the photo clearly shows searchlights focused on an alien spaceship; however, the photo was heavily modified by photo retouching prior to publication, a routine practice in graphic arts of the time intended to improve contrast in black and white photos. Los Angeles Times writer Larry Harnisch noted that the retouched photo along with faked newspaper headlines were presented as true historical material in trailers for the film *Battle: Los Angeles*. Harnisch commented, "if the publicity campaign wanted to establish UFO research as nothing but lies and fakery, it couldn't have done a better job."

1946 **Ghost rockets** (Swedish: *Spökraketer*, also called **Scandinavian ghost rockets**) were rocket- or missile-shaped unidentified flying objects sighted in 1946, mostly in Sweden and nearby countries.

The first reports of ghost rockets were made on February 26, 1946, by Finnish observers. About 2,000 sightings were logged between May and December 1946, with peaks on 9 and 11 August 1946. Two hundred sightings were verified with radar returns, and authorities recovered physical fragments which were attributed to ghost rockets.

Investigations concluded that many ghost rocket sightings were probably caused by meteors. For example, the peaks of the sightings, on the 9 and 11 August 1946, also fall within the peak of the annual Perseid meteor shower. However, most ghost rocket sightings did not occur during meteor shower activity, and furthermore displayed characteristics inconsistent with meteors, such as reported maneuverability.

Debate continues as to the origins of the unidentified ghost rockets. In 1946, however, it was thought likely that they originated from the former German rocket facility at Peenemünde, and were long-range tests by the Russians of captured German V-1 or V-2 missiles, or perhaps another early form of cruise missile because of the ways they were sometimes seen to maneuver. This prompted the Swedish Army to issue a directive stating that newspapers were not to report the exact location of ghost rocket sightings, or any information regarding the direction or speed of the object. This information, they reasoned, was vital for evaluation purposes to the nation or nations performing the tests.

The early Russian origins theory was rejected by Swedish, British, and U.S. military investigators because no recognizable rocket fragments were ever found, and according to some sightings the objects usually left no exhaust trail, some moved too slowly and usually flew horizontally, they sometimes traveled and maneuvered in formation, and they were usually silent. The sightings most often consisted of fast-flying rocket- or missile- shaped objects, with or without wings, visible for mere seconds. Instances of slower moving cigar shaped objects are also known. A hissing or rumbling sound was sometimes reported.

Crashes were not uncommon, almost always in lakes. Reports were made of objects crashing into a lake, then propelling themselves across the surface before sinking, as well as ordinary crashes.

The Swedish military performed several dives in the affected lakes shortly after the crashes, but found nothing other than occasional craters in the lake bottom or torn off aquatic plants.

The best known of these crashes occurred on July 19, 1946, into Lake Kölmjärv, Sweden. Witnesses reported a gray, rocket-shaped object with wings crashing in the lake. One witness interviewed heard a thunderclap, possibly the object exploding. However, a 3 week military search conducted in intense secrecy again turned up nothing.

Immediately after the investigation, the Swedish Air Force officer who led the search, Karl-Gösta Bartoll (photo right), submitted a report in which he stated that the bottom of the lake had been disturbed but nothing found and that "there are many indications that the Kölmjärv object disintegrated itself...the object was probably manufactured in a lightweight material, possibly a kind of magnesium alloy that would disintegrate easily, and not give indications on our instruments". When Bartoll was later interviewed in 1984 by Swedish researcher Clas Svahn, he again said their investigation suggested the object largely disintegrated in flight and insisted that "what people saw were real, physical objects".

On October 10, 1946, the Swedish Defense Staff publicly stated, "Most observations are vague and must be treated very skeptically. In some cases, however, clear, unambiguous observations have been made that cannot be explained as natural phenomena, Swedish aircraft, or imagination on the part of the observer. Echo, radar, and other equipment registered readings but gave no clue as to the nature of the objects". It was also stated that fragments alleged to have come from the missiles were nothing more than ordinary coke or slag.

On December 3, 1946, a memo was drafted for the Swedish Ghost Rocket committee stating "nearly one hundred impacts have been reported and thirty pieces of debris have been received and examined by FOA" (later said to be meteorite fragments). Of the nearly 1000 reports that had been received by the Swedish Defense Staff to November 29, 225 were considered observations of "real physical objects" and every one had been seen in broad daylight.

Michael Ryan

UFO-Memorial Ängelholm

05-18-1946 The **UFO-Memorial Ängelholm** is a shrine dedicated to a supposed UFO landing in Kronoskogen, a suburb of Ängelholm, Sweden. A few other such memorials exist in Europe (other examples include the memorial for the Robert Taylor incident in Livingston in Scotland and the Emilcin UFO memorial in Emilcin, Poland). Dedicated in 1963, it is situated in a forest clearing at Kronoskogen, which had witnessed numerous "large-scale test flights" in that time period. The UFO-Memorial Ängelholm memorialises the landing of a UFO, which is said to have taken place on 18 May 1946 and been seen by the Swedish entrepreneur, founder and owner of Cernelle AB, Gösta Carlsson. The memorial consists of a model of the UFO and the landing traces, and is constructed of concrete.

Clas Svahn, chairman of UFO-Sweden, has investigated the case and written a book together with Gösta Carlsson about the incident. According to him there was no convincing evidence that the event ever took place the way Gösta Carlsson described it.

06-21-1947 The **Maury Island Incident** refers to claims made by Fred Crisman and Harold Dahl of falling debris and threats by men in black following sightings of unidentified flying objects in the sky over Maury Island in Puget Sound in June 1947. Dahl later retracted his claims, stating the story was a hoax.

Crisman and Dahl said they were harbor patrolmen on a workboat who saw six doughnut shaped objects in the sky near Maury Island. According to Crisman and Dahl, one of the objects dropped a substance that resembled lava or "white metal" onto their boat, breaking a worker's arm and killing a dog. Dahl claimed he was later approached by a man in a dark suit and told not to talk about the incident. The story was later retold in Gray Barker's book "They Knew Too Much About Flying Saucers," which helped to popularize the image of "men in black" in mainstream culture.

The substance claimed to have been dropped by the objects was found to be slag from a local smelter. Years later, Dahl confessed to a reporter that the incident was a hoax.

PAGE 2   THE CHICAGO SUN, THURSDAY, JUNE 26, 1947

In These United States

Supersonic Flying Saucers Sighted by Idaho Pilot

Speed Estimated at 1,200 Miles an Hour When Seen 10,000 Feet Up Near Mt. Rainier

PENDLETON, Ore., June 25.—(P).

NINE bright, saucer-like objects flying at "incredible" speed at 10,000 feet altitude were reported here today by Kenneth Arnold, Boise (Idaho), pilot, who said he could not hazard a guess as to what they were.

Arnold, a U.S. Forest Service employee searching for a missing plane, said he sighted the mystery craft yesterday at 3 p.m. They were flying between Mount Rainier and Mount Adams, in Washington state, he said, and appeared to weave in and out of formation. Arnold said he clocked them and estimated their speed at 1,200 miles an hour.

Inquiries at Yakima last night brought only blank stares, he said, but he added he talked today with an unidentified man from Ukiah, south of here, who said he had seen similar objects over the mountains near Ukiah yesterday.

"It seems impossible," Arnold said, "but there it is."

• ♥ •

The **Kenneth Arnold UFO sighting** occurred on June 24, 1947, when private pilot Kenneth Arnold claimed that he saw a string of nine, shiny unidentified flying objects flying past Mount Rainier at speeds that Arnold estimated at a minimum of 1,200 miles an hour (1,932 km/hr). This was the first post-War sighting in the United States that garnered nationwide news coverage and is credited with being the first of the modern era of UFO sightings, including numerous reported sightings over the next two to three weeks. Arnold's description of the objects also led to the press quickly coining the terms *flying saucer* and *flying disc* as popular descriptive terms for UFOs.

On June 24, 1947, Arnold was flying from Chehalis, Washington to Yakima, Washington in a CallAir A-2 on a business trip. He made a brief detour after learning of a $5,000 reward for the discovery of a U.S. Marine Corps C-46 transport airplane that had crashed near Mt. Rainier. The skies were completely clear and there was a mild wind.

A few minutes before 3:00 p.m. at about 9,200 feet (2,800 m) in altitude and near Mineral, Washington, he gave up his search and started heading eastward towards Yakima. He saw a bright flashing light, similar to sun-light reflecting from a mirror. Afraid he might be dangerously close to another aircraft, Arnold scanned the skies around him, but all he could see was a DC-4 to his left and behind him, about 15 miles (24 km) away.

About 30 seconds after seeing the first flash of light, Arnold saw a series of bright flashes in the distance off to his left, or north of Mt. Rainier, which was then 20 to 25 miles (40 km) away. He thought they might be reflections on his airplane's windows, but a few quick tests (rocking his airplane from side to side, removing his eyeglasses, later rolling down his side window) ruled this out. The reflections came from flying objects.

They flew in a long chain, and Arnold for a moment considered they might be a flock of geese, but quickly ruled this out for a number of reasons, including the altitude, bright glint, and obviously very fast speed. He then thought they might be a new type of jet and started looking intently for a tail and was surprised that he couldn't find any. They quickly approached Rainier and then passed in front, usually appearing dark in profile against the bright white snowfield covering Rainier, but occasionally still giving off bright light flashes as they flipped around erratically. Sometimes he said he could see them on edge, when they seemed so thin and flat they were practically invisible. According to Jerome Clark, Arnold described them as a series of objects with convex shapes, though he later revealed that one object differed by being crescent-shaped. Several years later, Arnold would state he likened their movement to saucers skipping on water, without comparing their actual shapes to saucers, but initial quotes from him do indeed have him comparing the shape to like a "saucer", "disc", "pie pan", or "half moon", or generally convex and thin (discussion below). At one point Arnold said they flew behind a subpeak of Rainier and briefly disappeared. Knowing his position and the position of the (unspecified) subpeak, Arnold placed their distance as they flew past Rainier at about 23 miles (37 km).

Using a dzus cowling fastener as a gauge to compare the nine objects to the distant DC-4, Arnold estimated their angular size as slightly smaller than the DC-4, about the width between the outer engines (about 60 feet). Arnold also said he realized that the objects would have to be quite large to see any details at that distance and later, after comparing notes with a United Airlines crew that had a similar sighting 10 days later (see below), placed the absolute size as larger than a DC-4 airliner (or greater than 100 feet (30 m) in length).

Army Air Force analysts would later estimate 140 to 280 feet (85 m), based on analysis of human visual acuity and other sighting details (such as estimated distance). Arnold said the objects were grouped together, as Ted Bloecher writes, "in a diagonally stepped-down, echelon formation, stretched out over a distance that he later calculated to be five miles". Though moving on a more or less level horizontal plane, Arnold said the objects weaved from side to side ("like the tail of a Chinese kite" as he later stated), darting through the valleys and around the smaller mountain peaks. They would occasionally flip or bank on their edges in unison as they turned or maneuvered causing almost blindingly bright or mirror-like flashes of light. The encounter gave him an "eerie feeling", but Arnold suspected he had seen test flights of a new U.S. military aircraft. As the objects passed Mt Rainer, Arnold turned his plane southward on a more or less parallel course. It was at this point that he opened his side window and began observing the objects unobstructed by any glass that might have produced reflections. The objects did not disappear and continued to move very rapidly southward, continuously moving forward of his position. Curious about their speed, he began to time their rate of passage: he said they moved from Mt. Rainer to Mount Adams where they faded from view, a distance of about 50 miles (80 km), in one minute and forty-two seconds, according to the clock on his instrument panel. When he later had time to do the calculation, the speed was over 1,700 miles per hour (2,700 km/h). This was about three times faster than any manned aircraft in 1947. Not knowing exactly the distance where the objects faded from view, Arnold conservatively and arbitrarily rounded this down to 1,200 miles (1,900 km) an hour, still faster than any known aircraft, which had yet to break the sound barrier. It was this supersonic speed in addition to the unusual saucer or disk description that seemed to capture people's attention.

SAUCERS OVER TULSA ?

Arnold's sighting was partly corroborated by a prospector named Fred Johnson on Mt. Adams, who wrote AAF intelligence that he saw six of the objects on June 24 at about the same time as Arnold, which he viewed through a small telescope. He said they were "round" and tapered "sharply to a point in the head and in an oval shape." He also noted that the objects seemed to disturb his compass. An evaluation of the witness by AAF intelligence found him to be credible. Ironically, Johnson's report was listed as the first unexplained UFO report in Air Force files, while Arnold's was dismissed as a mirage, yet Johnson seemed to be describing a continuation of the same event as Arnold.

The Portland *Oregon Journal* reported on July 4 receiving a letter from an L. G. Bernier of Richland, Washington (about 110 miles (180 km) east of Mt. Adams and 140 miles (230 km) southeast of Mt. Rainier). Bernier wrote that he saw three of the strange objects over Richland flying "almost edgewise" toward Mt. Rainier about one half hour before Arnold. Bernier thought the three were part of a larger formation.

He indicated they were traveling at high speed: "I have seen a P-38 appear seemingly on one horizon and then gone to the opposite horizon in no time at all, but these disks certainly were traveling faster than any P-38. [Maximum speed of a P-38 was about 440 miles an hour.] No doubt Mr. Arnold saw them just a few minutes or seconds later, according to their speed." The previous day, Bernier had also spoken to his local newspaper, the Richland Washington *Villager*, and was among the first witnesses to suggest extraterrestrial origins: "I believe it may be a visitor from another planet."

About 60 miles (97 km) west-northwest of Richland in Yakima, Washington, a woman named Ethel Wheelhouse likewise reported sighting several flying discs moving at fantastic speeds at around the same time as Arnold's sighting.

When military intelligence began investigating Arnold's sighting in early July (see below), they found yet another witness from the area. A member of the Washington State forest service, who had been on fire watch at a tower in Diamond Gap, about 20 miles (32 km) south of Yakima, reported seeing "flashes" at 3:00 p.m. on the 24th over Mount Rainier (or exactly the same time as Arnold's sighting), that appeared to move in a straight line. Similarly, at 3:00 p.m. Sidney B. Gallagher in Washington state (exact position unspecified) reported seeing nine shiny discs flash by to the north.

A Seattle newspaper also mentioned a woman near Tacoma who said she saw a chain of nine, bright objects flying at high speed near Mt. Rainier. Unfortunately this short news item wasn't precise as to time or date, but indicated it was around the same date as Arnold's sighting.

However, a pilot of a DC-4 some 10 to 15 miles (24 km) north of Arnold en route to Seattle reported seeing nothing unusual. (This was the same DC-4 seen by Arnold and which he used for size comparison.)

Other Seattle area newspapers also reported other sightings of flashing, rapidly moving unknown objects on the same day, but not the same time, as Arnold's sighting. Most of these sightings were over Seattle or west of Seattle in the town of Bremerton, either that morning or at night. Altogether, there were at least 16 other reported UFO sightings the same day as Arnold's in the Washington state area.

The primary corroborative sighting, however, occurred ten days later (July 4) when a United Airlines crew over Idaho en route to Seattle also spotted five to nine disk-like objects that paced their plane for 10 to 15 minutes before suddenly disappearing. The next day in Seattle, Arnold met with the pilot, Cpt. Emil J. Smith, and copilot and compared sighting details. The main difference in shape was that the United crew thought the objects appeared rough on top. This was one of the few sightings that Arnold felt was reliable, most of the rest he thought were the public seeing other things and letting their imaginations run wild. Arnold and Cpt. Smith became friends, met again with Army Air Force intelligence officers on July 12 and filed sighting reports, then teamed up again at the end of July in investigating the strange Maury Island incident.

A similar sighting of eight objects also occurred over Tulsa, Oklahoma on July 12, 1947. In this instance, a photo was taken and published in the Tulsa *Daily World* the following day (photo at right). Interestingly, the photographer, Enlo Gilmore, said that in blowups of the photo, the objects resembled baseball catcher's mitts or flying wings. He was of the opinion that the military had a secret fleet of flying wing airplanes.

He had been a gunnery officer in the Navy during the war, and using information from another witness, also a veteran, he performed a triangulation and arrived at an estimation of speed of 1,700 miles per hour (2,700 km/h), or essentially the same estimate as Arnold's. One of the objects, he said, seemed to have a hole in the middle.

Two or three photos of a similar, solitary object were taken by William Rhodes over Phoenix, Arizona on July 7, 1947, and appeared in a local Phoenix newspaper and some other newspapers. The object was rounded in front with a crescent back. These photos also seem to show something resembling a hole in the middle, though Rhodes thought it was a canopy. Rhodes's negatives and prints were later confiscated by the FBI and military. However, the photos show up in later Air Force intelligence reports.

Arnold was soon shown the Rhodes photos when he met with two AAF intelligence officers. He commented, "It was a disk almost identical to the one peculiar flying saucer that had been worrying me since my original observation — the one that looked different from the rest and that I had never mentioned to anyone." As a result, Arnold felt that the Rhodes photos were genuine.

07-1947 In mid 1947, an object crashed near a ranch near Roswell, New Mexico, prompting claims alleging the crash was of an extraterrestrial spaceship. The U.S. Air force initially reported a flying disc had crashed.

After an initial spike of interest, the military reported that the crash was merely of a conventional weather balloon. Interest subsequently waned until the late 1970s when ufologists began promulgating a variety of increasingly elaborate conspiracy theories, claiming that one or more alien spacecraft had crash-landed, and that the extraterrestrial occupants had been recovered by the military who then engaged in a cover-up.

In the 1990s, the US military published reports disclosing the true nature of the crashed Project Mogul balloon.

Nevertheless, the Roswell incident continues to be of interest in popular media, and conspiracy theories surrounding the event persist. Roswell has been called "the world's most famous, most exhaustively investigated UFO claim".

The sequence of events that was triggered by the crash of the object near Roswell. On July 8, 1947, the Roswell Army Air Field

(RAAF) public information officer Walter Haut, issued a press release stating that personnel from the field's 509th Operations Group had recovered a "flying disc", which had crashed on a ranch near Roswell.

The military rebutted this statement and stated a weather balloon had crashed with a nuclear monitoring device attached. A press conference was held, featuring debris (foil, rubber and wood) said to be from the crashed object, which matched the weather balloon description. Historian Robert Goldberg wrote that the intended effect was achieved: "the story died the next day".

Subsequently the incident faded from the attention of UFO enthusiasts for more than 30 years.

On June 14, 1947, William Brazel, a foreman working on the Foster homestead, noticed clusters of debris approximately 30 miles (50 km) north of Roswell, New Mexico. This date—or "about three weeks" before July 8—appeared in later stories featuring Brazel, but the initial press release from the Roswell Army Air Field (RAAF) said the find was "sometime last week", suggesting Brazel found the debris in early July. Brazel told the *Roswell Daily Record* that he and his son saw a "large area of bright wreckage" He paid little attention to it but returned on July 4 with his son, wife and daughter to gather up the material. Some accounts have described Brazel as having gathered some of the material earlier, rolling it together and stashing it under some brush. The next day, Brazel heard reports about "flying discs" and wondered if that was what he had picked up. On July 7, Brazel saw Sheriff Wilcox and "whispered kinda confidential like" that he may have found a flying disc. Another account quotes Wilcox as saying Brazel reported the object on July 6.

Wilcox called RAAF Major Jesse Marcel and a "man in plain clothes" accompanied Brazel back to the ranch where more pieces were picked up. "[We] spent a couple of hours Monday afternoon [July 7] looking for any more parts of the weather device", said Marcel. "We found a few more patches of tinfoil and rubber."

As described in the July 9, 1947 edition of the *Roswell Daily Record*,

The balloon which held it up, if that was how it worked, must have been 12 feet long, [Brazel] felt, measuring the distance by the size of the room in which he sat. The rubber like material was smoky gray in color and scattered over an area about 200 yards in diameter. When the debris was gathered up, material made a bundle about three feet long and 7 or 8 inches thick, while the rubber like material made a bundle about 18 or 20 inches long and about 8 inches thick. In all, he estimated, the entire lot would have weighed maybe five pounds. There was no sign of any metal in the area which might have been used for an engine, and no sign of any propellers of any kind.

There were no words to be found anywhere on the instrument, although there were symbols on some of the parts.

A piece of metal was found which when crumpled up in ones have would subsequently unfold again, back into its original shape without creases (as if it was trying to heal).

A telex sent to a Federal Bureau of Investigation (FBI) office from the Fort Worth, Texas office quoted a Major from the Eighth Air Force (also based in Fort Worth at Carswell Air Force Base) on July 8, 1947 as saying that "The disc is hexagonal in shape and was suspended from a balloon by cable, which balloon was approximately twenty feet in diameter. Major Curtan further advices that the object found resembles a high altitude weather balloon with a radar reflector, but that telephonic conversation between their office and Wright field had not [UNINTELLIGIBLE] borne out this belief."

 A National Oceanic and Atmospheric Administration (NOAA) weather balloon after launching

Early on Tuesday, July 8, the RAAF issued a press release, which was immediately picked up by numerous news outlets:

The many rumors regarding the flying disc became a reality yesterday when the intelligence office of the 509th Bomb group of the Eighth Air Force, Roswell Army Air Field, was fortunate enough to gain possession of a disc through the cooperation of one of the local ranchers and the sheriff's office of Chaves County. The flying object landed on a ranch near Roswell sometime last week. Not having phone facilities, the rancher stored the disc until such time as he was able to contact the sheriff's office, who in turn notified Maj. Jesse A. Marcel of the 509th Bomb Group Intelligence Office. Action was immediately taken and the disc was picked up at the rancher's home. It was inspected at the Roswell Army Air Field and subsequently loaned by Major Marcel to higher headquarters.

Colonel William H. Blanchard, commanding officer of the 509th, contacted General Roger M. Ramey of the Eighth Air Force in Fort Worth, Texas, and Ramey ordered the object be flown to Fort Worth Army Air Field. At the base, Warrant Officer Irving Newton confirmed Ramey's preliminary opinion, identifying the object as being a weather balloon and its "kite", a nickname for a radar reflector used to track the balloons from the ground. Another news release was issued, this time from the Fort Worth base, describing the object as being a "weather balloon".

Between 1978 and the early 1990s, UFO researchers such as Stanton T. Friedman, William Moore, Karl T. Pflock, and the team of Kevin D. Randle and Donald R. Schmitt interviewed several hundred people who had – or claimed to have had – a connection with the events at Roswell in 1947.

Hundreds of documents were obtained via Freedom of Information Act requests, and some were supposedly leaked by insiders, such as the so-called Majestic 12. Their conclusions were at least one alien craft had crashed in the Roswell vicinity, aliens – some possibly still alive – had been recovered, and a government cover-up of any knowledge of the incident had taken place.

Over the years, books, articles, television specials, and a made-for-TV movie brought the 1947 incident significant notoriety. By the mid-1990s, public polls such as a 1997 CNN/*Time* poll, revealed that the majority of people interviewed believed that aliens had indeed visited Earth, and that aliens had landed at Roswell, but that all the relevant information was being kept secret by the US government.

According to anthropologists Susan Harding and Kathleen Stewart, the Roswell Story was the prime example of how a discourse moved from the fringes to the mainstream according to the prevailing *zeitgeist*: public preoccupation in the 1980s with "conspiracy, cover-up and repression" aligned well with the Roswell narratives as told in the "sensational books" which were being published.

In 1978, nuclear physicist and author Stanton T. Friedman interviewed Jesse Marcel, the only person known to have accompanied the Roswell debris from where it was recovered to Fort Worth where reporters saw material which was claimed to be part of the recovered object. The accounts given by Friedman and others in the following years elevated Roswell from a forgotten incident to perhaps the most famous UFO case of all time.

1948 **Green fireballs** are a type of unidentified flying object which have been sighted in the sky since the late 1940s. Early sightings primarily occurred in the southwestern United States, particularly in New Mexico. They were once of notable concern to the US government because they were often clustered around sensitive research and military installations, such as Los Alamos and Sandia National Laboratory, then Sandia base.

Meteor expert Dr. Lincoln LaPaz headed much of the investigation into the fireballs on behalf of the military. LaPaz's conclusion was that the objects displayed too many anomalous characteristics to be a type of meteor and instead were artificial, perhaps secret Russian spy devices. The green fireballs were seen by many people of high repute including LaPaz, distinguished Los Alamos scientists, Kirtland AFB intelligence officers and Air Command Defense personnel. A February 1949 Los Alamos conference attended by aforementioned sighters, Project Sign, world-renowned upper atmosphere physicist Dr. Joseph Kaplan, H-bomb scientist Dr. Edward Teller, other scientists and military brass concluded, though far from unanimously, that green fireballs were natural phenomena. To the conference attendees, though the green fire ball source was unknown, their existence was unquestioned. Secret conferences were convened at Los Alamos to study the phenomenon and in Washington by the U.S. Air Force Scientific Advisory Board.

In December 1949 Project Twinkle, a network of green fireball observation and photographic stations, was established but never fully implemented. It was discontinued two years later, with the official conclusion that the phenomenon was probably natural in origin.

Green fireballs have been given natural, man-made, and extraterrestrial origins and have become associated with both the Cold War and ufology. Because of the extensive government paper trail on the phenomenon, many ufologists consider the green fireballs to be among the best documented examples of unidentified flying objects (UFOs).

Some early reports came from late November 1948, but were at first dismissed as military green flares. Then on the night of December 5, 1948, two separate plane crews, one military (Air Force C-47, Captain Goede, 9:27 p.m., 10 miles (16 km) east of Albuquerque) and one civilian (DC-3, Pioneer Flight 63, 9:35 p.m., east of Las Vegas, New Mexico), each asserted that they had seen a "green ball of fire"; the C-47 crew had seen an identical object 22 minutes before near Las Vegas. The military crew described the light as like a huge green meteor except it arched upwards and then flat instead of downwards The civilian crew described the light as having a trajectory too low and flat for a meteor, at first abreast and ahead of them but then appearing to come straight at them on a collision course, forcing the pilot to swerve the plane at which time the object appeared full moon size.

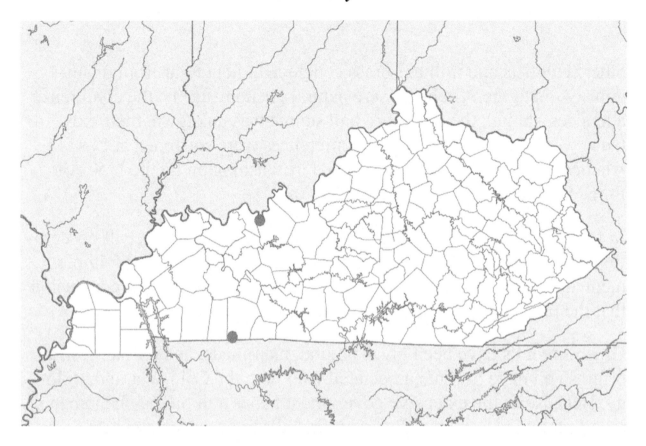

01-07-1948 The **Mantell UFO incident** was among the most publicized early UFO reports. The incident resulted in the crash and death of 25-year-old Kentucky Air National Guard pilot, Captain Thomas F. Mantell, on January 7, 1948 while in pursuit of a UFO.

Historian David Michael Jacobs argues the Mantell case marked a sharp shift in both public and governmental perceptions of UFOs. Previously, the news media often treated UFO reports with a whimsical or glib attitude reserved for silly season news. Following Mantell's death, however, Jacobs notes "the fact that a person had died in an encounter with an alleged flying saucer dramatically increased public concern about the phenomenon. Now a dramatic new prospect entered thought about UFOs: they might be not only extraterrestrial but potentially hostile as well."

However, later investigation by the US Air Force's Project Blue Book indicated that Mantell died chasing a Skyhook balloon, which in 1948 was a top-secret project that Mantell would not have known about.

Mantell was an experienced pilot; his flight history consisted of 2,167 hours in the air, and he had been honored for his part in the Battle of Normandy during World War II.

On 7 January 1948, Godman Field at Fort Knox, Kentucky received a report from the Kentucky Highway Patrol of an unusual aerial object near Maysville, Kentucky. Reports of a westbound circular object, 250 feet (76 m) to 300 feet (91 m) in diameter, were received from Owensboro, Kentucky, and Irvington, Kentucky.

At about 1:45 p.m., Sergeant Quinton Blackwell saw an object from his position in the control tower at Fort Knox. Two other witnesses in the tower also reported a white object in the distance. Colonel Guy Hix, the base commander, reported an object he described as "very white," and "about one fourth the size of the full moon ... Through binoculars it appeared to have a red border at the bottom ... It remained stationary, seemingly, for one and a half hours." Observers at Clinton County Army Air Field in Ohio described the object "as having the appearance of a flaming red cone trailing a gaseous green mist" and observed the object for around 35 minutes. Another observer at Lockbourne Army Air Field in Ohio noted, "Just before leaving it came to very near the ground, staying down for about ten seconds, then climbed at a very fast rate back to its original altitude, 10,000 feet, leveling off and disappearing into the overcast heading 120 degrees. Its speed was greater than 500 mph in level flight."

01-07-1948 The **Chiles-Whitted UFO encounter** occurred on July 24, 1948 in the skies near Montgomery, Alabama. Two commercial pilots, Clarence S. Chiles and John B. Whitted, claimed that at approximately 2:45 AM on July 24 they observed a "glowing object" pass by their plane before it appeared to pull up into a cloud and travel out of sight. According to Air Force officer and Project Blue Book supervisor Edward J. Ruppelt, the Chiles-Whitted sighting was one of three "classic" UFO incidents in 1948 that convinced the personnel of Project Sign, Blue Book's predecessor, "that UFOs were real." However, later studies by Air Force and civilian researchers indicated that Chiles and Whitted had seen a meteor, possibly a bolide, and in 1959 Project Blue Book formally stated that a meteor was the cause of the incident.

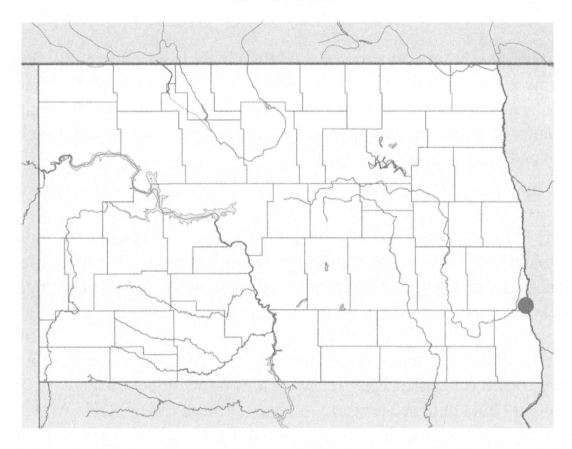

10-01-1948 The **Gorman UFO dogfight** was a widely publicized UFO incident. It occurred on October 1, 1948, in the skies over Fargo, North Dakota, and involved George F. Gorman, a pilot with the North Dakota National Guard. USAF Captain Edward J. Ruppelt wrote in his bestselling and influential *The Report on Unidentified Flying Objects* that the Gorman Dogfight was one of three "classic" UFO incidents in 1948 that "proved to [Air Force] intelligence specialists that UFOs were real." However, in 1949 the US Air Force labeled the Gorman Dogfight as being caused by a lighted weather balloon.

Gorman told the tower that he was going to pursue the object to determine its identity. He moved his Mustang to full power (350 to 400 MPH), but soon realized that the object was going too fast for him to catch it in a straight run. Instead, he tried cutting the object off by turns.

He made a right turn and approached the object head-on at 5,000 feet; the object flew over his plane at a distance of about 500 feet. Gorman described the object as a simple "ball of light" about six to eight inches in diameter. He also noted later that when the object increased its speed, it stopped blinking and grew brighter.

After his near-collision, Gorman lost sight of the object; when he saw it again it appeared to have made a 180-degree turn and was coming at him again. The object then made a sudden vertical climb; Gorman followed the object in his own steep climb. At 14,000 feet his P-51 stalled; the object was still 2,000 feet above him. Gorman made two further attempts to get closer to the object, with no success. It seemed to make another head-on pass, but broke off before coming close to his fighter. By this point the object had moved over the Fargo Airport, in the control tower the air traffic controller, L.D. Jensen, viewed the object through binoculars but could see no form or shape around the light. He was joined by Cannon and his passenger from the Piper Cub; they had landed and walked to the control tower to get a better view of the object.

Gorman continued to follow the object until he was approximately 25 miles southwest of Fargo. At 14,000 feet he observed the light at 11,000 feet; he then dived on the object at full power. However, the object made a vertical climb. Gorman tried to pursue but watched as the object passed out of visual range. At this point he broke off the chase; it was 9:27 PM. He flew back to Fargo's Hector Airport.

On October 23, 1948, Gorman gave a sworn account of the incident to investigators. His statement was often reprinted in future years in numerous books and documentaries about UFOs. The statement read:

I am convinced that there was definite thought behind its maneuvers.

I am further convinced that the object was governed by the laws of inertia because its acceleration was rapid but not immediate and although it was able to turn fairly tight at considerable speed, it still followed a natural curve. When I attempted to turn with the object I blacked out temporarily due to excessive speed. I am in fairly good physical condition and I do not believe that there are many if any pilots who could withstand the turn and speed effected by the object, and remain conscious. The object was not only able to out turn and out speed my aircraft...but was able to attain a far steeper climb and was able to maintain a constant rate of climb far in excess of my aircraft.

The Gorman Dogfight received wide national publicity and helped fuel the wave of UFO reports in the late forties. Although some UFO researchers, such as Dr. James E. McDonald, a physicist at the University of Arizona, and retired US Marine Corps Major Donald Keyhoe, disagreed with the Air Force's conclusions and continued to regard the case as unsolved, other UFO researchers agreed with Project Sign's conclusions in the case. As UFO historian Jerome Clark writes, "unlike some Air Force would-be solutions this one seems plausible" and that, in his opinion, "After the Mantell Incident the Gorman sighting may be the most overrated UFO report in the early history of the phenomenon."

10-15-1948 Fukuoka Incident. The radar of a F-61 Black Widow detected a target below the aircraft. While the aircrew tried to intercept it, the pilot saw the object, which appeared as a stubby cigar; then the object accelerated and disappeared.

05-11-1950 The **McMinnville UFO photographs** were taken on a farm near McMinnville, Oregon, in 1950. The photos were reprinted in *Life* magazine and in newspapers across the nation, and are often considered to be among the most famous ever taken of a UFO. The photos remain controversial, with many ufologists claiming they show a genuine, unidentified object in the sky, while many UFO skeptics claim that the photos are a hoax.

At 7:30 p.m. on May 11, 1950, Evelyn Trent was walking back to her farmhouse after feeding rabbits on her farm. Mrs. Trent and her husband Paul lived on a farm approximately nine miles from McMinnville (the Trent farm was actually located just outside of Sheridan, Oregon). Before reaching the house she claimed to see a "slow-moving, metallic disk-shaped object heading in her direction from the northeast."

She yelled for her husband, who was inside the house; upon leaving the house he claimed he also saw the object. After a short time he went back inside their home to obtain a camera; he managed to take two photos of the object before it sped away to the west. Paul Trent's father claimed he briefly viewed the object before it flew away.

It took some time for Paul Trent to have the film developed, and he apparently sought no publicity immediately following the incident. When he mentioned the incident to his banker, Frank Wortmann, the banker was intrigued enough to display the photos from his bank window in McMinnville. Shortly afterwards Bill Powell, a local reporter, convinced Mr. Trent to loan him the negatives. Powell examined the negatives and found no evidence that they were tampered with or faked. On June 8, 1950, Powell's story of the incident—accompanied by the two photos—was published as a front-page story in the local McMinnville newspaper, the *Telephone-Register*. The headline read: "At Long Last—Authentic Photographs Of Flying Saucer[?]" The story and photos were subsequently picked up by the International News Service (INS) and sent to other newspapers around the nation, thus giving them wide publicity. *Life* magazine published cropped versions of the photos on June 26, 1950, along with a photo of Trent and his camera. The Trents had been promised that the negatives would be returned to them; however, they were not returned—*Life* magazine told the Trents that it had misplaced the negatives.

In 1967 the negatives were found in the files of the United Press International (UPI), a news service which had merged with INS years earlier. The negatives were then loaned to Dr. William K. Hartmann, an astronomer who was working as an investigator for the Condon Committee, a government-funded UFO research project based at the University of Colorado Boulder.

The Trents were not immediately informed that their "lost" negatives had been found.

Hartmann interviewed the Trents and was impressed by their sincerity; the Trents never received any money for their photos, and he could find no evidence that they had sought any fame or fortune from them. In Hartmann's analysis, he wrote to the Condon Committee that "This is one of the few UFO reports in which all factors investigated, geometric, psychological, and physical, appear to be consistent with the assertion that an extraordinary flying object, silvery, metallic, disk-shaped, tens of meters in diameter, and evidently artificial, flew within sight of two witnesses."

After Hartmann concluded his investigation he returned the negatives to UPI, which then informed the Trents about them. In 1970 the Trents asked Philip Bladine, the editor of the *News-Register* (the successor of the *Telephone-Register*), for the negatives; the Trents noted that they had never been paid for the negatives and thus wanted them back. Bladine asked UPI to return the negatives, which it did. However, for some reason Bladine never told the Trents that the negatives had been returned. In 1975 the negatives were found in the files of the *News-Register* by Dr. Bruce Maccabee, an optical physicist for the U.S. Navy and a ufologist. Maccabee did his own extensive analysis of the negatives and concluded that they were not hoaxed and showed a "real, physical object" in the sky above the Trent's farm. He then ensured that the negatives were finally returned to the Trents.

In the 1980s two UFO skeptics, Philip J. Klass and Robert Sheaffer, argued that the photos were faked, and that the entire event was a hoax.

Their primary argument was that shadows on a garage in the left-hand side of the photos proved that the photos were taken in the morning rather than in the early evening, as the Trents had claimed. Klass and Sheaffer argued that since the Trents had apparently lied about the time the photos were taken, their entire story was thus suspect. Klass and Sheaffer also argued that the Trents had shown an interest in UFOs prior to their sighting, and their analysis of the photos indicated that the object photographed was small and likely a model hanging from power lines visible at the top of the photos. They also believed the object may have been the detached rear-view mirror of a vehicle. When Sheaffer sent his studies on the case to William Hartmann, Hartmann withdrew the positive assessment of the case he had sent to the Condon Committee. However, Maccabee offered a rebuttal to the Klass-Sheaffer theory by arguing that cloud conditions in the McMinnville area on the evening of the sighting could have caused the shadows, and that a close analysis of the UFO indicated that it was not suspended from the power lines and was in fact located some distance above the Trent's farm; thus, in his opinion, the Klass-Sheaffer explanation was flawed

In April 2013, three researchers with IPACO posted two studies to their website entitled "Back to McMinnville pictures" and "Evidence of a suspension thread." They used proprietary computer software designed to analyze UFO photos by François Louange, who previously has done image analysis for NASA, the European Space Agency, and GEIPAN. They concluded the geometry of the photographs is most consistent with a small model with a hollow bottom hanging from a wire, though no thread was detected. Further investigation, however, detected the presence of a thread, and the study concluded: "the clear result of this study was that the McMinnville UFO was a model hanging from a thread."

In August 2013, UFO researcher Brad Sparks posted a rebuttal regarding the IPACO McMinnville UFO studies in which he stated that the studies contained multiple foundation measurement inconsistencies. An example of this was IPACO applying and working with a 5 inch diameter for the UFO [in the McMinnville photos] in some instances, and also applying and working with a 6 inch diameter for it in others. Sparks argued that certain measurements within the studies were manipulated whenever they proved unable to support the "UFO-model-hanging-by-a-string" hoax hypothesis.

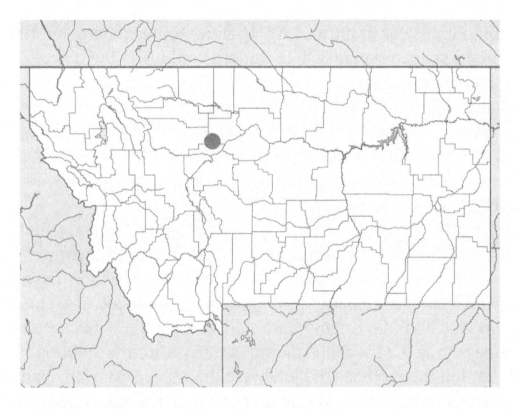

08-05-1950 The **Mariana UFO Incident** occurred in August 1950 in Great Falls, Montana. The film footage of the sighting is believed to be among the first ever taken of what came to be called an unidentified flying object, and was investigated by the US Air Force.

At 11:25 am on August 15, 1950, Nick Mariana, the general manager of the Great Falls Electrics minor-league baseball team, and his nineteen-year-old secretary, Virginia Raunig, were inspecting the empty Legion Stadium baseball field before a game. The Electrics were a farm club of the Brooklyn Dodgers. A bright flash caught Mariana's eye and, according to his reports, he saw two bright silvery objects, rotating while flying over Great Falls at a speed he estimated to be two hundred to four hundred miles per hour. He believed that they were roughly fifty feet wide and one hundred and fifty feet apart.

Mariana ran to his car to retrieve his 16 mm movie camera and filmed the UFOs for sixteen seconds. The camera could film the objects in color, but could not record sound. Raunig also witnessed the objects. The day after Mariana's sighting, the *Great Falls Tribune*, the city's daily newspaper, described his sighting and the film in an article, which was picked up by other media outlets. For several weeks after the sighting, Mariana showed his film to local community groups, including the Central Roundtable Athletic Club.

After seeing the film, a reporter for the *Great Falls Tribune* called Wright-Patterson Air Force Base in Ohio and informed them of Mariana's sighting and film. U.S. Air Force Captain John P. Brynildsen interviewed Mariana at nearby Malmstrom AFB outside of Great Falls. When Mariana and Ms. Raunig both told him that they had seen two jet fighters pass over the baseball stadium shortly after the sighting, Brynildsen felt that perhaps the jets were the objects Mariana had seen and captured on film. With Mariana's permission, Capt. Brynildsen sent the film to Wright-Patterson AFB for analysis. He told a reporter in Great Falls that he had "picked up about eight feet of film from Mariana." However, in his message to Wright-Patterson he said that he was sending "approximately fifteen feet of moving picture film" to the base for study. According to UFO historian Jerome Clark, this discrepancy was never cleared up.

At Wright-Patterson AFB the film was briefly examined and determined to be the reflections from two F-94 jet fighters that were known to be flying over Great Falls at the time of Mariana's sighting. Lt. Col. Ray W. Taylor returned the film to Mariana with a cover letter stating that "our photoanalysts were unable to find anything identifiable of an unusual nature".

However, according to Air Force officer Edward J. Ruppelt, who would become the supervisor of the Air Force's Project Blue Book investigation into the UFO mystery in 1951, "in 1950 there was no interest [by the Air Force] in the UFO, so after a quick viewing, Project Grudge had written them off as the reflections from two F-94 jet fighters that were in the area". Controversy soon arose when Mariana claimed that the first thirty-five frames of his film - which he said most clearly showed the UFOs as rotating disks - were missing. People in the Great Falls area who had viewed Mariana's film supported his claims.They claimed that the missing frames clearly showed the UFOs as spinning, metallic disks with a "notch or band" along their outer edges. The Air Force personnel denied this accusation, and insisted that they had removed only a single frame of film which was damaged in the analysis. The controversy over the "missing frames" was never resolved.

08-25-1951 The **Lubbock Lights,** were an unusual formation of lights seen over the city of Lubbock, Texas, from August–September 1951. The Lubbock Lights incident received national publicity and is regarded as one of the first great UFO cases in the United States. The Lubbock Lights were investigated by the US Air Force in 1951, which initially believed they were caused by a type of bird called a plover, but eventually concluded that the lights "weren't birds... but they weren't spaceships...

the [Lubbock Lights] have been positively identified as a very commonplace and easily explainable natural phenomenon." However, to maintain the anonymity of the scientist who had provided the explanation, the Air Force refrained from providing any details regarding their explanation for the lights.

The first publicized sighting of the lights occurred on August 25, 1951, at around 9 pm. Three professors from Texas Technological College (now Texas Tech University), located in Lubbock, were sitting in the backyard of one of the professor's homes when they observed the "lights" fly overhead. A total of 20-30 lights, as bright as stars but larger in size, flew over the yard in a matter of seconds. The professors immediately ruled out meteors as a possible cause for the sightings, and as they discussed their sighting a second, similar, group of lights flew overhead. The three professors - Dr. A.G. Oberg, chemical engineer, Dr. W.L. Ducker, a department head and petroleum engineer, and Dr. W.I. Robinson, a geologist - reported their sighting to the local newspaper, the *Lubbock Avalanche-Journal*. Following the newspaper's article, three women in Lubbock reported that they had observed "peculiar flashing lights" in the sky on the same night of the professor's sightings. Dr. Carl Hemminger, a professor of German at Texas Tech, also reported seeing the objects, as did the head of the college's journalism department.The three professors became determined to view the objects again and perhaps discover their identity. On September 5, 1951, all three men, along with two other professors from Texas Tech, were sitting in Dr. Robinson's frontyard when the lights flew overhead. According to Dr. Grayson Mead the lights "appeared to be about the size of a dinner plate and they were greenish-blue, slightly fluorescent in color. They were smaller than the full moon at the horizon.

There were about a dozen to fifteen of these lights... they were absolutely circular... it gave all of us... an extremely eerie feeling." Mead claimed that the lights could not have been birds, but he also stated that they "went over so fast... that we wished we could have had a better look." The professors observed one formation of lights flying above a thin cloud at about 2,000 feet (610 m); this allowed them to calculate that the lights were traveling at over 600 miles per hour (970 km/h).

On the evening of August 30, 1951, Carl Hart, Jr., a freshman at Texas Tech, was lying in bed looking out of the window of his room when he observed a group of 18-20 white lights in a "v" formation flying overhead. Hart took a 35-mm Kodak camera and walked to the backyard of his parent's home to see if the lights would return. Two more flights passed overhead, and Hart was able to take a total of five photos before they disappeared. After having the photos developed Hart took them to the offices of the *Lubbock Avalanche-Journal*. After examining the photos the newspaper's editor, Jay Harris, told Hart that he would print them in the paper, but that he would "run him (Hart) out of town" if the photos were fake. When Hart assured him that the photos were genuine, Harris paid Hart $10 for the pictures. The photographs were soon reprinted in newspapers around the nation, and were printed in *Life* magazine, thus giving them wide publicity. The physics laboratory at Wright-Patterson Air Force Base in Ohio analyzed the Hart photographs. After an extensive analysis and investigation of the photos, Lieutenant Edward J. Ruppelt, the supervisor of the Air Force's Project Blue Book, released a written statement to the press that "the [Hart] photos were never proven to be a hoax, but neither were they proven to be genuine." Hart has consistently maintained to this day that the photos are genuine. Curiously, the Texas Tech professors claimed that the photos did not represent what they had seen, since their objects had flown in a "u" formation instead of the "v" formation depicted in Hart's photos.

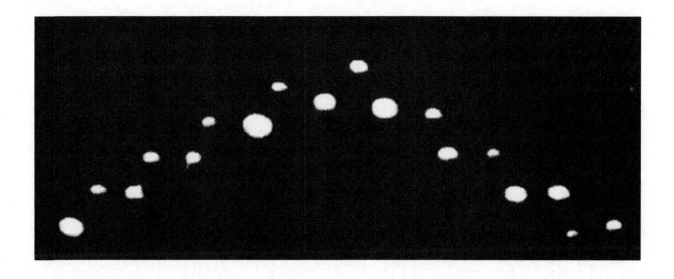

09-10-1951 The **Lubbock Lights** were an unusual formation of lights seen over the city of Lubbock, Texas, from August–September 1951. The Lubbock Lights incident received national publicity and is regarded as one of the first great UFO cases in the United States. The Lubbock Lights were investigated by the US Air Force in 1951, which initially believed they were caused by a type of bird called a plover, but eventually concluded that the lights "weren't birds... but they weren't spaceships...the [Lubbock Lights] have been positively identified as a very commonplace and easily explainable natural phenomenon." However, to maintain the anonymity of the scientist who had provided the explanation, the Air Force refrained from providing any details regarding their explanation for the lights.

The first publicized sighting of the lights occurred on August 25, 1951, at around 9 pm. Three professors from Texas Technological College (now Texas Tech University), located in Lubbock, were sitting in the backyard of one of the professor's homes when they observed the "lights" fly overhead. A total of 20-30 lights, as bright as stars but larger in size, flew over the yard in a matter of seconds. The professors immediately ruled out meteors as a possible cause for the sightings, and as they discussed their sighting a second, similar, group of lights flew overhead.

The three professors - Dr. A.G. Oberg, chemical engineer, Dr. W.L. Ducker, a department head and petroleum engineer, and Dr. W.I. Robinson, a geologist - reported their sighting to the local newspaper, the *Lubbock Avalanche-Journal*. Following the newspaper's article, three women in Lubbock reported that they had observed "peculiar flashing lights" in the sky on the same night of the professor's sightings. Dr. Carl Hemminger, a professor of German at Texas Tech, also reported seeing the objects, as did the head of the college's journalism department.

The three professors became determined to view the objects again and perhaps discover their identity. On September 5, 1951, all three men, along with two other professors from Texas Tech, were sitting in Dr. Robinson's frontyard when the lights flew overhead. According to Dr. Grayson Mead the lights "appeared to be about the size of a dinner plate and they were greenish-blue, slightly fluorescent in color. They were smaller than the full moon at the horizon. There were about a dozen to fifteen of these lights... they were absolutely circular... it gave all of us... an extremely eerie feeling." Mead claimed that the lights could not have been birds, but he also stated that they "went over so fast... that we wished we could have had a better look." The professors observed one formation of lights flying above a thin cloud at about 2,000 feet (610 m); this allowed them to calculate that the lights were traveling at over 600 miles per hour (970 km/h).

09-10-1951 Fort Monmouth UFO Case. At Fort Monmouth Base, a radar operator picked up an unknown target. Seventeen minutes later, the crew of a T-33 saw a discus-shaped unidentified object.

07-12-1952 The **1952 Washington, D.C. UFO incident**, also known as the **Washington flap** or the **Washington National Airport Sightings**, was a series of unidentified flying object reports from July 12 to July 29, 1952, over Washington, D.C. The most publicized sightings took place on consecutive weekends, July 19–20 and July 26–27.

At 11:40 p.m. on Saturday, July 19, 1952, Edward Nugent, an air traffic controller at Washington National Airport (today Ronald Reagan Washington National Airport), spotted seven objects on his radar. The objects were located 15 miles (24 km) south-southwest of the city; no known aircraft were in the area and the objects were not following any established flight paths. Nugent's superior, Harry Barnes, a senior air-traffic controller at the airport watched the objects on Nugent's radarscope. He later wrote:

We knew immediately that a very strange situation existed . . . their movements were completely radical compared to those of ordinary aircraft.

Barnes had two controllers check Nugent's radar; they found that it was working normally. Barnes then called National Airport's other radar center; the controller there, Howard Cocklin, told Barnes that he also had the objects on his radarscope. Furthermore, Cocklin said that by looking out of the control tower window he could see one of the objects, "a bright orange light. I can't tell what's behind it."

At this point, other objects appeared in all sectors of the radarscope; when they moved over the White House and the United States Capitol, Barnes called Andrews Air Force Base, located 10 miles from National Airport. Although Andrews reported that they had no unusual objects on their radar, an airman soon called the base's control tower to report the sighting of a strange object. Airman William Brady, who was in the tower, then saw an "object which appeared to be like an orange ball of fire, trailing a tail . . . [it was] unlike anything I had ever seen before." As Brady tried to alert the other personnel in the tower, the strange object "took off at an unbelievable speed." Meanwhile, another person in the National Airport control tower reported seeing "an orange disk about 3,000 feet altitude." On one of the airport's runways, S.C. Pierman, a Capital Airlines pilot, was waiting in the cockpit of his DC-4 for permission to take off. After spotting what he believed to be a meteor, he was told that the control tower's radar had picked up unknown objects closing in on his position. Pierman observed six objects — "white, tailless, fast-moving lights" — over a 14-minute period. Pierman was in radio contact with Barnes during his sighting, and Barnes later related that "each sighting coincided with a pip we could see near his plane.

When he reported that the light streaked off at a high speed, it disappeared on our scope."

At Andrews Air Force Base, meanwhile, the control tower personnel were tracking on radar what some thought to be unknown objects, but others suspected, and in one instance were able to prove, were simply stars and meteors. However, Staff Sgt. Charles Davenport observed an orange-red light to the south; the light "would appear to stand still, then make an abrupt change in direction and altitude . . . this happened several times." At one point both radar centers at National Airport and the radar at Andrews Air Force Base were tracking an object hovering over a radio beacon. The object vanished in all three radar centers at the same time. At 3 a.m., shortly before two United States Air Force F-94 Starfire jet fighters from New Castle Air Force Base in Delaware arrived over Washington, all of the objects vanished from the radar at National Airport. However, when the jets ran low on fuel and left, the objects returned, which convinced Barnes that "the UFOs were monitoring radio traffic and behaving accordingly." The objects were last detected by radar at 5:30 a.m. Around sunrise, E.W. Chambers, a civilian radio engineer in Washington's suburbs, claimed to observe "five huge disks circling in a loose formation. They tilted upward and left on a steep ascent."

07-16-1952 Salem, Massachusetts UFO incident. Coast Guard photographer Shell Alpert took a photograph of four roughly elliptical blobs of light in formation through the window of his photographic laboratory. A Coast Guard press release described the lights as "objects", however Seaman Alpert subsequently issued a statement saying, "I cannot in all honesty say that I saw objects or aircraft, merely some manner of lights."

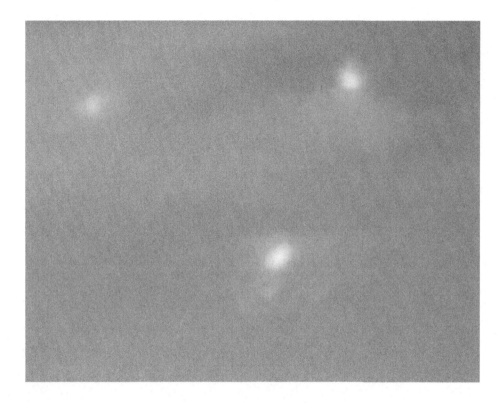

07-24-1952 Two pilots saw three unusual Delta wing aircraft flying in a V-formation over Carson Sink.

The **Carson Sink Case** is a UFO incident alleged to have taken place near Carson Sink in western Nevada in the United States on July 24, 1952.

Edward Ruppelt claims that two U.S. Air Force Colonels requisitioned a twin engine B-25 bomber at Hamilton Field north of San Francisco for a cross-country flight to Colorado. At 3:40 P.M. MST while at 11,000 feet (3300 m) over the Carson Sink area east of Reno, according to Ruppelt, the two pilots saw three unknown aircraft make a left bank and fly quickly to within 400 to 800 yards (meters) of their B-25.

The two men estimated the speed of the unknown aircraft to be at the very least three times that of the F-86. After four seconds,

the aircraft sped away out of the vision of the pilots. When they landed in Colorado Springs, they contacted Air Defense Command Headquarters and learned that no civilian or military aircraft had been anywhere near the Carson Sink at the time of the incident. The two men dismissed the suggestion that they had seen F-86 jets, since they were intimately familiar with the design of that craft. Air Defense Command relayed the report to Ruppelt at Project Blue Book. In his subsequent book, *The Report on Unidentified Flying Objects*, Ruppelt characterized it as a "good UFO report with an unknown conclusion".

Braxton Co. Residents Faint, Become Ill After Run-In With Weird 10-Foot Monster

Seven Braxton County residents Saturday reporting seeing a 10-foot Frankenstein-like monster in the hills above Flatwood.

They said they saw the monster Friday night when they climbed a wooded hill to investigate reports that a flying saucer had landed.

Mrs. Kathlyn May, Flatwood, said she and six boys, including a 17-year-old member of the National Guard, started to search for a bright object which her two small sons said they had seen come down.

However, State Police laughed the reports off as hysteria. They said the so-called monster had grown from seven to 17 feet in 24 hours.

The National Guard member, Gene Lemon, was leading the group when he said he saw what appeared to be a pair of bright eyes in a tree. At first he thought it was an opossum or a raccoon but when he shone his flashlight on it, he said, he saw a 10-foot monster with a blood-red face and a green body that seemed to glow.

Mrs. May said Lemon let out a terrified scream and fell over backwards. She said the monster started toward them with a bounding motion.

All of the party agreed that there was an overpowering smell that burned the nostrils and made them sick. Several of the party fainted and vomited for several hours after returning to town.

A. Lee Stewart, co-publisher of the Braxton County Democrat, said he and several men armed with shotguns returned with Lemon about a half-hour to an hour later, and reported a sickening odor still present. He said there were also slight heat waves in the air.

"Those people were the most scared people I've ever seen," Stewart said. "People don't make up that kind of story that quickly."

Both Mrs. May and Lemon described the thing as having the shape of a man, blood-red face, bright green body, protruding eyes, and hand extended forward and appeared to give off an eerie light. They said it had a black shield affair in the shape of an ace of spades behind it and wore what looked like a pleated metallic shirt.

"It looked worse than Frankenstein," Mrs. May said.

The **Flatwoods Monster**, also known as the **Braxton County Monster** or the **Phantom of Flatwoods**, is an alleged unidentified extraterrestrial or cryptid reported to have been sighted in the town of Flatwoods in Braxton County, West Virginia, United States, on September 12, 1952. Stories of the creature are an example of a purported close encounter of the third kind.

Various descriptions of the entity exist. Most agree that it was at least 7 feet (2.1 m) tall, with a black body and a dark, glowing face. Witnesses described the creature's head as elongated, shaped like a sideways diamond, and as having non-human eyes; a large, circular cowling appeared behind the head. The creature's body was *inhumanly-shaped* and clad in a dark pleated exoskeleton, later described as a shadow.

Some accounts record that the creature appeared to have "no visible arms" due to its incredible speed.

Others reported long, stringy arms, protruding from the front of its body, with long, claw-like fingers. The monster is referred to as the "Lizard Monster" on the March 10, 2010, episode of *MonsterQuest*. A large, pulsating red ball of light that hovered above or rested on the ground was associated with the monster. Ufologists believe that it may have been a powered craft that the entity had piloted.

At 7:15 p.m. on September 12, 1952, two brothers, Edward and Fred May, and their friend Tommy Hyer (ages 13, 12, and 10 respectively) witnessed a bright object cross the sky. The object appeared to come to rest on land belonging to local farmer G. Bailey Fisher. Upon witnessing the object, the boys went to the home of the May brothers' mother, Kathleen May, where they reported seeing a UFO crash land in the hills. From there, Mrs. May, accompanied by the three boys, local children Neil Nunley (14) and Ronnie Shaver (10), and 17-year-old West Virginia National Guardsman Eugene 'Gene' Lemon, traveled to the Fisher farm in an effort to locate whatever it was that the boys had seen.

Lemon's dog ran ahead out of sight and suddenly began barking, and moments later ran back to the group with its tail between its legs. After traveling about 0.25 miles (402 m) the group reached the top of a hill, where they reportedly saw a large pulsating "ball of fire" about 50 feet (15 m) to their right. They also detected a pungent mist that made their eyes and noses burn. Lemon then noticed two small lights over to the left of the object, underneath a nearby oak tree and directed his flashlight towards them, revealing the creature, which was reported to have emitted a shrill hissing noise before gliding towards them, changing direction and then heading off towards the red light.

At this point the group fled in panic. Upon returning home, Mrs. May contacted local Sheriff Robert Carr and Mr. A. Lee Stewert, co-owner of the *Braxton Democrat*, a local newspaper. Stewert conducted a number of interviews and returned to the site with Lemon later that night, where he reported that "there was a sickening, burnt, metallic odor still prevailing". Sheriff Carr and his deputy Burnell Long searched the area separately, but reported finding no trace of the encounter other than the smell. Early the next morning, Stewert visited the site of the encounter for a second time and discovered two elongated tracks in the mud, as well as traces of a thick black liquid. He immediately reported them as being possible signs of a saucer landing, based on the premise that the area had not been subjected to vehicle traffic for at least a year. It was later revealed that the tracks were likely to have been those of a 1942 Chevrolet pickup truck driven by local Max Lockard, who had gone to the site to look for the creature some hours prior to Stewert's discovery.

After the event, Mr. William and Donna Smith, investigators associated with Civilian Saucer Investigation, LA, obtained a number of accounts from witnesses who claimed to have experienced a similar or related phenomena. These accounts included the story of a mother and her 21-year-old-daughter, who claimed to have encountered a creature with the same appearance and odor a week prior to the September 12 incident. The encounter reportedly affected the daughter so badly that she was confined to Clarksburg Hospital for three weeks. They also gathered a statement from the mother of Eugene Lemon, in which she said that, at the approximate time of the crash, her house had been violently shaken and her radio had cut out for 45 minutes, and a report from the director of the local Board of Education in which he claimed to have seen a flying saucer taking off at 6:30 a.m. on September 13 (the morning after the creature was sighted).

05-21-1953 Prescott Sightings Three Prescott residents sight a total of eight craft at Del Rio Springs Creek, 20 miles north of Prescott.

08-12-1953 On August 5 and August 6, 1953 the US Military investigated a UFO incident in Bismarck, North Dakota. What has become known as the Ellsworth Case is one of the most significant radar-visual cases in the annals of UFO sightings.

The event was witnessed by almost forty-five agitated citizens along with military Air Defense System personnel. The object was first sighted by Miss Kellian at 8:00 P.M. on August 5. The description was of a red glowing light making long sweeping movements.

11-23-1953 The disappearance of Felix Moncla and Robert Wilson. U.S. Air Force pilot and radar operator and their F-89C disappeared while pursuing an unidentified radar return.

Felix Eugene Moncla, Jr. and Robert Wilson presumably died November 23, 1953) were United States Air Force pilots who mysteriously disappeared while performing an air defense intercept over Lake Superior in 1953. This is sometimes known as **The Kinross Incident**, after Kinross Air Force Base, where Moncla was on temporary assignment when they disappeared.

The U.S. Air Force reported that Moncla had crashed and that the object of the intercept was a Royal Canadian Air Force aircraft.

According to the report, the pilot of the Canadian aircraft was later contacted and reported that he did not see the F-89 and did not know that he was the subject of an interception.

On multiple occasions, the RCAF denied that any of their aircraft were "involved in any incident" on that day, in correspondence with members of the public asking for further details on the intercept.

On the evening of November 23, 1953, Air Defense Command Ground Intercept radar operators at Sault Ste. Marie, Michigan identified an unusual target near the Soo Locks. An F-89C Scorpion jet from Kinross Air Force Base was scrambled to investigate the radar return; the Scorpion was piloted by First Lieutenant Moncla with Second Lieutenant Robert L. Wilson acting as the Scorpion's radar operator.

Wilson had problems tracking the object on the Scorpion's radar, so ground radar operators gave Moncla directions towards the object as he flew. Moncla eventually closed in on the object at about 8000 feet in altitude.

Ground Control tracked the Scorpion and the unidentified object as two "blips" on the radar screen. The two blips on the radar screen grew closer and closer, until they seemed to merge as one (return). Assuming that Moncla had flown either under or over the target, Ground Control thought that moments later, the Scorpion and the object would again appear as two separate blips. Donald Keyhoe reported that there was a fear that the two objects had struck one another "as if in a smashing collision."

Rather, the single blip continued on its previous course.

Attempts were made to contact Moncla via radio, but this was unsuccessful. A search and rescue operation by both the USAF and the RCAF was quickly mounted, but failed to find a trace of the plane or the pilots. Weather conditions were a factor hampering the search.

The official USAF Accident Investigation Report states the F-89 was sent to investigate an RCAF C-47 Skytrain which was travelling off course.

The F-89 was flying at an elevation of 8000 feet when it merged with the other aircraft, as was expected in an interception. Its IFF signal also disappeared after the two returns merged on the radar scope. Although efforts to contact the crew on radio were unsuccessful, the pilot of another F-89 sent on the search stated in testimony to the accident board that he believed that he had heard a brief radio transmission from the pilot about forty minutes after the plane disappeared.

Air Force investigators reported that Moncla may have experienced vertigo and crashed into the lake. The Air Force said that Moncla had been known to experience vertigo from time to time: "Additional leads uncovered during this later course of the investigation indicated that there might be a possibility that Lt. Moncla was subjective to attacks of vertigo in a little more than the normal degree. Upon pursuing these leads, it was discovered that statements had been made by former members of Lt. Moncla's organization but were not first hand evidence and were regarded as hearsay." Pilot vertigo is not listed as a cause or possible cause in any of the USAF Accident Investigation Board's findings and conclusions.

12-16-1953 Kelly Johnson/Santa Barbara Channel Case. Legendary Lockheed aircraft engineer Clarence "Kelly" Johnson, designer of the F-104, U-2, and SR-71, and his wife observed a huge Flying Wing over the Pacific from the ground in Agoura. Meanwhile, one of Johnson's flight test crews aboard an WV-2 (see EC-121) spotted the craft from Long Beach, California. USAF concluded these trained observers had seen a lenticular cloud, even though Johnson considered and ruled out that explanation.

The **Kelly–Hopkinsville encounter**, also known as the **Hopkinsville Goblins Case**, and to a lesser extent the **Kelly Green Men Case**, was a series of incidents of alleged close encounters with extraterrestrial beings. These were reported in 1955, the most famous and well-publicized of which centered on a rural farmhouse, at the time belonging to the Sutton family, which was located between the hamlet of Kelly and the small city of Hopkinsville, both in Christian County, Kentucky, United States.

Members of two families at the farmhouse reported seeing unidentifiable creatures. Other witnesses attested to lights in the sky and odd sounds.

The events are regarded by UFOlogists as one of the most significant and well-documented cases in the history of UFO incidents, and are a favorite for study in ufology. UFOlogists have claimed it was investigated by the United States Air Force, although no evidence of an investigation has been found.

Witnesses to the incidents included eleven people belonging to the two families present at the farmhouse and others in the area. Several local policemen and a state trooper saw unexplained lights in the night sky and heard odd noises the same night.

The seven people present in the farmhouse claimed that they were terrorized by an unknown number of creatures similar to gremlins, which have since often been referred to as the "Hopkinsville Goblins" in popular culture. The residents of the farmhouse described them as being around three feet tall, with upright pointed ears, thin limbs (their legs were said to be almost in a state of atrophy), long arms and claw-like hands or talons. The creatures were either silvery in color or wearing something metallic. On occasion, their movements seemed to defy gravity, with them floating above the ground and appearing in high-up places. They "walked" with a swaying motion as though wading through water. Although the creatures never entered the house, they would pop up at windows and at the doorway, waking up the children in the house in a hysterical frenzy. The families fled the farmhouse in the middle of the night to the local police station and sheriff Russell Greenwell noted they were visibly shaken. The families returned to the farmhouse with Sheriff Greenwell and twenty officers, yet the occurrences continued. Police saw evidence of the struggle and damage to the house, as well as seeing strange lights and hearing noises themselves. The witnesses additionally claimed to have used firearms to shoot at the creatures with little or no effect. In addition, the house and surrounding grounds were extensively damaged during the incident.

UFO researcher Allan Hendry wrote "[t]his case is distinguished by its duration and also by the number of witnesses involved." Jerome Clark writes that "investigations by police, Air Force officers from nearby Fort Campbell, and civilian ufologists found no evidence of a hoax". Author Brian Dunning contends that "The claim that Air Force investigators showed up the next day at Mrs. Lankford's house has been published a number of times by later authors, but I could find no corroborating evidence of this." Dunning also observes that "the four military police who accompanied the police officers on the night of the event were from an Army base, not an Air Force base." Project Blue Book listed the case as a hoax with no further comment.

07-24-1956 Drakensberg Contactee. A well-known photo series depicting a supposed UFO, was taken on 24 July near Rosetta in the Drakensberg region. The photographer, meteorologist Elizabeth Klarer, claimed detailed adventures with an alien race, besides having had an alien lover, Akon, who would have fathered her son Ayling.

05-03-1957 Edwards Air Force Base flying saucer filming. In 1957, when Gordon Cooper was 30 and a Captain, he was assigned to Fighter Section of the Experimental Flight Test Engineering Division at Edwards AFB in California. He acted as a test pilot and project manager. On May 3 of that year, he had a crew setting up an Askania Cinetheodolite precision landing system on a dry lake bed. This cinetheodolite system would take pictures at one frame per second as an aircraft landed. The crew consisted of James Bittick and Jack Gettys who began work at the site just before 0800, using both still and motion picture cameras. According to his accounts, later that morning they returned to report to Cooper that they saw a "strange-looking, saucer-like" aircraft that did not make a sound either on landing or take-off.

According to his accounts, Cooper realized that these men, who on a regular basis have seen experimental aircraft flying and landing around them as part of their job of filming those aircraft, were clearly worked up and unnerved. They explained how the saucer hovered over them, landed 50 yards away from them using three extended landing gears and then took off as they approached for a closer look. Being photographers with cameras in hand, they of course shot images with 35mm and 4×5 still cameras as well as motion picture film. There was a special Pentagon number to call to report incidents like this. He called and it immediately went up the chain of command until he was instructed by a general to have the film developed (but to make no prints of it) and send it right away in a locked courier pouch. As he had not been instructed to *not* look at the negatives before sending them, he *did*. He said the quality of the photography was excellent as would be expected from the experienced photographers who took them. What he saw was exactly what they had described to him. He did not see the movie film before everything was sent away. He expected that there would be a follow up investigation since an aircraft of unknown origin had landed in a highly classified military installation, but nothing was ever said of the incident again. He was never able to track down what happened to those photos. He assumed that they ended up going to the Air Force's official UFO investigation, Project Blue Book, which was based at Wright-Patterson Air Force Base.

He held claim until his death that the U.S. government was indeed covering up information about UFOs. He gave the example of President Harry Truman who he claimed said on April 4, 1950,

"I can assure you that flying saucers, given that they exist, are not constructed by any power on Earth." He also pointed out that there were hundreds of reports made by his fellow pilots, many coming from military jet pilots sent to respond to radar or visual sightings from the ground. In his memoirs, Cooper wrote he had seen other unexplained aircraft several times during his career, and also said hundreds of similar reports had been made. He further claimed these sightings had been "swept under the rug" by the U.S. government. Throughout his later life Cooper expressed repeatedly in interviews he had seen UFOs and described his recollections for the documentary *Out of the Blue*.

05-20-1957 Milton Torres 1957 UFO Encounter. U.S. Air Force fighter pilot Milton Torres reports that he was ordered to intercept and fire on a UFO displaying "very unusual flight patterns" over East Anglia. Ground radar operators had tracked the object for some time before Torres' plane was scrambled to intercept.

09-1957 Ubatuba UFO Explosion. Two fishermen watched a UFO crash and explosion, and retrieved fragments of the object.

09-04-1957 Portugal "mothership" sighting. The crews of 4 fighter-bombers watched a large luminescent UFO and smaller attendant UFOs during a nighttime flight at 25,000 ft.

10-16-1957 Antonio Vilas Boas Abduction. At the time of his alleged abduction, Antônio Vilas-Boas was a 23-year-old Brazilian farmer who was working at night to avoid the hot temperatures of the day. On October 16, 1957, he was ploughing fields near São Francisco de Sales when he saw what he described as a "red star" in the night sky. According to his story, this "star" approached his position, growing in size until it became recognizable as a roughly circular or egg-shaped aerial craft, with a red light at its front and a rotating cupola on top. The craft began descending to land in the field, extending three "legs" as it did so. At that point, Boas decided to run from the scene. According to Boas, he first attempted to leave the scene on his tractor, but when its lights and engine died after traveling only a short distance, he decided to continue on foot. However, he was seized by a 1.5 m (five-foot) tall humanoid, who was wearing grey coveralls and a helmet.

Its eyes were small and blue, and instead of speech it made noises like barks or yelps. Three similar beings then joined the first in subduing Boas, and they dragged him inside their craft. Once inside the craft, Boas said that he was stripped of his clothes and covered from head-to-toe with a strange gel. He was then led into a large semicircular room, through a doorway that had strange red symbols written over it. (Boas claimed that he was able to memorize these symbols and later reproduced them for investigators.) In this room the beings took samples of Boas' blood from his chin. After this he was then taken to a third room and left alone for around half an hour. During this time, some kind of gas was pumped into the room, which made Boas become violently ill.

Shortly after this, Boas claimed that he was joined in the room by another humanoid. This one, however, was female, very attractive, and naked. She was the same height as the other beings he had encountered, with a small, pointed chin and large, blue catlike eyes. The hair on her head was long and white (somewhat like platinum blonde) but her underarm and pubic hair were bright red. Boas said he was strongly attracted to the woman, and the two had sexual intercourse. During this act, Boas noted that the female did not kiss him but instead nipped him on the chin.

When it was all over, the female smiled at Boas, rubbing her belly and gestured upwards. Boas took this to mean that she was going to raise their child in space. The female seemed relieved that their "task" was over, and Boas himself said that he felt angered by the situation, because he felt as though he had been little more than "a good stallion" for the humanoids. Boas said that he was then given back his clothing and taken on a tour of the ship by the humanoids.

During this tour he said that he attempted to take a clock-like device as proof of his encounter, but was caught by the humanoids and prevented from doing so. He was then escorted off the ship and watched as it took off, glowing brightly. When Boas returned home, he discovered that four hours had passed.

Antonio Vilas Boas later became a lawyer, married and had four children. He stuck to the story of his alleged abduction for his entire life. Though some sources say he died in 1992, he died on January 17, 1991.

Following this alleged event, Boas claimed to have suffered from nausea and weakness, as well as headaches and lesions on the skin which appeared with any kind of light bruising. Eventually, he contacted journalist Jose Martins, who had placed an ad in a newspaper looking for people who had had experiences with UFOs. Upon hearing Boas' story, Martins contacted Dr. Olavo Fontes of National School of Medicine of Brazil; Fontes was also in contact with the American UFO research group APRO. Fontes examined the farmer and concluded that he had been exposed to a large dose of radiation from some source and was now suffering from mild radiation sickness. Writer Terry Melanson states:

> Among [Boas's] symptoms were 'pains throughout the body, nausea, headaches, loss of appetite, ceaselessly burning sensations in the eyes, cutaneous lesions at the slightest of light bruising...which went on appearing for months, looking like small reddish nodules, harder than the skin around them and protuberant, painful when touched, each with a small central orifice yielding a yellowish thin waterish discharge.' The skin surrounding the wounds presented 'a hyperchromatic violet-tinged area.'

According to Researcher Peter Rogerson, the story first came to light in February, 1958, and the earliest definite print reference to Boas' story was from the April–June 1962 issue of the Brazilian UFO periodical *SBESDV Bulletin*. Rogerson notes that the story had definitely circulated between 1958 and 1962, and was probably recorded in print, but that details are uncertain.

Boas was able to recall every detail of his purported experience without the need for hypnotic regression. Further, Boas' experience occurred in 1957, which was still several years before the famous Hill abduction which made the concept of alien abduction famous and opened the door to many other reports of similar experiences.

Researcher Peter Rogerson, however, doubts the veracity of Boas' story. He notes that several months before Boas first related his claims, a similar story was printed in the November 1957 issue of the periodical *O Cruzeiro*, and suggests that Boas borrowed details of this earlier account, along with elements of the contactee stories of George Adamski. Rogerson also argues:

> One reason why the [Boas] story gained credibility was the prejudiced assumption that any farmer in the Brazilian interior had to be an illiterate peasant who 'couldn't make this up'. As Eddie Bullard pointed out to me, the fact that the Villas Boas family possessed a tractor put them well above the peasant class ... We now know that AVB was a determinedly upwardly mobile young man, studying a correspondence course and eventually becoming a lawyer (at which news the ufologists who had considered him too much the rural simpleton to have made the story up, now argued that he was too respectable and bourgeois to have done so).

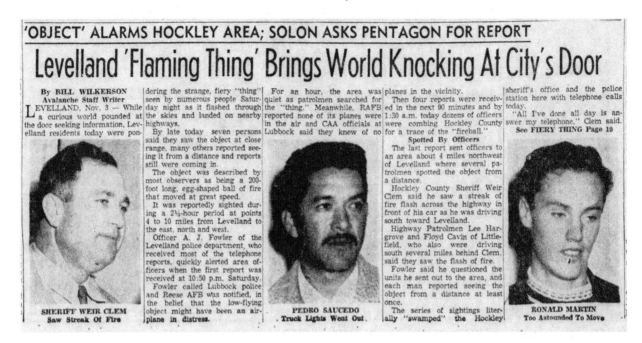

See FIERY THING Page 10

11-02-57 Levelland UFO Case. The incident began late on the evening of November 2 when two immigrant farm workers, Pedro Saucedo and Joe Salaz, called the Levelland police department to report a UFO sighting. Saucedo told police officer A.J. Fowler, who was working the night desk at the police station, that they had been driving four miles (6 km) west of Levelland when they saw a blue flash of light near the road. They claimed their truck's engine died, and a rocket-shaped object rose up and approached the truck. According to Saucedo, "I jumped out of the truck and hit the dirt because I was afraid. I called to Joe but he didn't get out. The thing passed directly over my truck with a great sound and rush of wind. It sounded like thunder and my truck rocked from the flash...I felt a lot of heat." As the object moved away the truck's engine restarted and worked normally. Believing the story to be a joke, Fowler ignored it. An hour later, motorist Jim Wheeler reported a "brilliantly lit, egg-shaped object, about 200 feet long" was sitting in the road, four miles (6 km) east of Levelland, blocking his path. He claimed his vehicle died and as he got out of his car the object took off and its lights went out. As it moved away,

Wheeler's car restarted and worked normally. At 10:55 pm a married couple driving northeast of Levelland reported that they saw a bright flash of light moving across the sky and their headlights and radio died for three seconds. Five minutes later Jose Alvarez claimed he met the strange object sitting on the road 11 miles (18 km) north of Levelland, and his vehicle's engine died until the object departed. At 12:05 am (November 3), a Texas Technological College (now Texas Tech University) student named Newell Wright was surprised when, driving 10 miles (16 km) east of Levelland, his "car engine began to sputter, the ammeter on the dash jumped to discharge and then back to normal, and the motor started cutting out like it was out of gas...the car rolled to a stop; then the headlights dimmed and several seconds later went out." When he got out to check on the problem, he saw a "100-foot-long" egg-shaped object sitting in the road. It took off, and his engine started running again. At 12:15 am Officer Fowler received another call, this time from a farmer named Frank Williams who claimed he had encountered a brightly glowing object sitting in the road, and "as his car approached it, its lights went out and its motor stopped." The object flew away, and his car's lights and motor started working again. Other callers were Ronald Martin at 12:45 am and James Long at 1:15 am, and they both reported seeing a brightly lit object sitting in the road in front of them, and they also claimed that their engines and headlights died until the object flew away. By this time, several Levelland police officers were actively investigating the incident. Among them was Sheriff Weir Clem, who saw a brilliant red object moving across the sky at 1:30 am. At 1:45 am Levelland's Fire Chief, Ray Jones, also saw the object and his vehicle's lights and engine sputtered. The sightings apparently ended soon after this incident. During the night of November 2–3, the Levelland police department received a total of 15 phone calls concerning the strange object, and Officer Fowler noted that "everybody who called was very excited."

01-16-1958 Trindade UFO Incident. At 12:00 p.m. on January 16, 1958, Brazilian ship Almirante Saldanha, taking part in projects of the International Geophysical Year, was preparing to sail away from Ilha de Trindade, off the coast of the state of Espírito Santo.

Among the crew of the ship, there was a member of the Brazilian Air Force, Captain José Teobaldo Viegas, the submarine photographer Almiro Baraúna, several scientists and a group of highly trained explorers.

Reportedly Captain Viegas was on the deck with several scientists and members of the crew when he suddenly noticed a flying object, which had a "ring" around it, just like Saturn.

Witnesses present reportedly saw the U.F.O. at the same time. It reportedly came toward the island from the east, flew towards the "Pico Desejado" (*Wished Peak*), made a steep turn and went away very quickly to the north-west. As soon as the object was noticed, Almiro Baraúna was sought for photography. After getting the camera and going up the quarter-deck, he managed to take several pictures of the object.

After being disclosed, the pictures were exhaustively analyzed by the *Laboratório de Reconhecimento Aéreo da Marinha* (Brazilian Navy's Aerial Reconnaissance Laboratory) and by the *Serviço Aerofotogramétrico Cruzeiro do Sul*. All the pictures were considered genuine, taken from a real occurrence, with the certification of military personnel. The pictures even received confirmation from Juscelino Kubitschek, president of Brazil at that time. To this day because of the certification of the witnesses and the official recognition, the object in Almiro Baraúna's pictures remains unexplained.

In 2010, a document containing a testimonial by Baraúna was unveiled:

> On January 16, 1958, the Navy training ship "Almirante Saldanha" was moored at a bay in Trindade Island, some eight hundred miles off the coast of Espírito Santo. It was around 11:00 a.m., the sky was clear and the crew was getting ready to return to Rio de Janeiro when suddenly a group of people at the stern, among them the retired Força Aérea Brasileira captain-aviator José Viegas, alerted everyone.

> Instantly, all who were in the deck (some fifty people) started to see a strange, silver, saucer-shaped object moving above the sea towards the island. The object didn't make any sound, was shiny and sometimes it moved quickly, then slowly, up and slightly down and when it accelerated,

it would leave a white phosphorescent trail that would disappear short-ly. In its trajectory, the object disappeared behind the Desejado Peak and all expected it to reappear on the other side of the island, it reap-peared from the same direction, stopped for some seconds and then dis-appeared again with great speed at the horizon. At first, when the ob-ject returned, I was able to take six pictures, two of which were lost due to the confusion at the deck, and the other four showed the object on the horizon, in a reasonable sequence, approaching the island from the mountain's side, and finally disappearing, going away. I took the film from my camera twenty minutes later following the commander's request, who wanted to know if the pictures had good quality. Almost the entire crew of the ship saw the film and they were unanimous in their reports to the Secret Service of the Brazilian Navy. These were the crew of the ship:

Chief Amilar Vieira Filho, banker, diver and athlete; Vice-chief: re-tired Força Aérea Brasileira captain-aviator José Viegas; Divers: Aluizio and Mauro; Photographer: Almiro Baraúna. The group above was part of a group of spearfishing from Icaraí. Among the five mem-bers, only Mauro and Aluizio didn't see the object because they were at the ship's kitchen and when they ran to see it, it was already gone. Ac-cording to the rumors I heard at the deck, the electronic equipment of the ship stopped working during the apparition of the object; what I can confirm is that after the ship left the island, the equipment malfunc-tioned three more times and the officials didn't have any plausible ex-planation for what was happening. Every time the ship stopped, the lights weakened slowly until the point they completely went out. When this happened, the officials would walk to the deck with their binocu-lars, however, the sky was clouded and they couldn't see anything.

I need to say that if the reporter of the newspaper *Correio da Manhã* hadn't been smart enough to take copies of the pictures offered to then president Juscelino Kubitscheck, maybe no one would ever be aware of this pictures since the Navy had "marked" me, asking how much did I want not to publicize the pictures. I would like to make it clear that every official with whom I had contact during all the period of the interrogation were quite lovely with me, I felt completely comfortable and they didn't impose any objection to the possible unveiling of the case. they only mentioned that the sensationalist nature of the case could cause panic among the population and that was the reason the Brazilian Air Force wanted to avoid the publicizing of such cases

Other investigators of the case have questioned the authenticity of the photos. Project Blue Book (the controversial U.S. Air Force investigation into the U.F.O. phenomenon) concluded that the photographs were hoaxed. The credibility of Barauna himself has been questioned. He had produced hoaxed photographs in the past (not only of U.F.O.s) and in the past had written an article showing how a well-known U.F.O. photograph taken some years earlier could have been hoaxed. Also, Barauna had the negatives for two days before the Brazilian Navy took them from him for their investigation, and he had cut them away from the remainder of the film negatives. Also, the Brazilian Navy did not get statements from the witnesses immediately after the event. Therefore, the actual number of witnesses is not known with certainty.

02-02-1959 The **Dyatlov Pass incident** refers to the demise of nine hikers in mysterious circumstances on the night of February 2, 1959 in the northern Ural Mountains. The name *Dyatlov Pass* refers to the name of the group's leader, Igor Dyatlov.

The incident involved a group of ten from Ural Polytechnical Institute (Уральский политехнический институт, УПИ) who had set up camp for the night on the slopes of Kholat Syakhl. Investigators later determined that the skiers had torn their tents from the inside out.

They fled the campsite, some of them barefoot, under heavy snowfall. Although the bodies showed no signs of struggle, such as contusions, two victims had fractured skulls and broken ribs. Soviet authorities determined that an "unknown compelling force" had caused the deaths; access to the region was consequently blocked for hikers and adventurers for three years after the incident. Due to the lack of survivors, the chronology of events remains uncertain, although several explanations have been put forward, including a possible avalanche, a military accident, or a hostile encounter with a yeti or other unknown creature.

06-26-1959 Father William Booth Gill sighting. Missionary and many natives saw several UFOs, one of them seeming to be repaired by four human-like occupants; witnesses and aliens waved at each other. The case was investigated by J. A. Hynek and accompanied by other sightings.

Barney and Betty Hill were an American couple who were allegedly abducted by extraterrestrials in a rural portion of New Hampshire on September 19, 1961 through September 20, 1961.

The incident came to be called the "Hill Abduction" or the "Zeta Reticuli Incident" because the couple stated they had been kidnapped for a short time by a UFO.

It was the first widely publicized report of alien abduction, adapted into the best-selling 1966 book *The Interrupted Journey* and the 1975 television movie *The UFO Incident.*

Most of Betty Hill's notes, tapes, and other items have been placed in the permanent collection at the University of New Hampshire, her alma mater. In July 2011, the state Division of Historical Resources marked the site of the alleged craft's first approach with a historical marker.

According to a variety of reports given by the Hills, the alleged UFO sighting happened on September 19, 1961, around 10:30 p.m. The Hills were driving back to Portsmouth from a vacation in Niagara Falls and Montreal. There were only a few other cars on the road as they made their way home to New Hampshire's seacoast. Just south of Lancaster, New Hampshire, Betty claimed to have observed a bright point of light in the sky that moved from below the moon and the planet Jupiter, upward to the west of the moon. While Barney navigated U.S. Route 3, Betty reasoned that she was observing a falling star, only it moved upward. Since it moved erratically and grew bigger and brighter, Betty urged Barney to stop the car for a closer look, as well as to walk their dog, Delsey. Barney stopped at a scenic picnic area just south of Twin Mountain. Worried about the presence of bears, Barney retrieved a pistol that he kept in the trunk of the car.

Betty, through binoculars, observed an "odd-shaped" craft flashing multi-colored lights travel across the face of the moon. Because her sister had confided to her about having a flying saucer sighting several years earlier, Betty thought it might be what she was observing. Through binoculars Barney observed what he reasoned was a commercial airliner traveling toward Vermont on its way to Montreal. However, he soon changed his mind, because without looking as if it had turned, the craft rapidly descended in his direction. This observation caused Barney to realize, "this object that was a plane was *not* a plane." He quickly returned to the car and drove toward Franconia Notch, a narrow, mountainous stretch of the road.

The Hills claimed that they continued driving on the isolated road, moving very slowly through Franconia Notch in order to observe the object as it came even closer. At one point, the object passed above a restaurant and signal tower on top of Cannon Mountain. It passed over the mountain and came out near the Old Man of the Mountain. Betty testified that it was at least one and a half times the length of the granite cliff profile, which was 40 feet (12 m) long, and that seemed to be rotating. The couple watched as the silent, illuminated craft moved erratically and bounced back and forth in the night sky. As they drove along Route 3 through Franconia Notch, they stated that it seemed to be playing a game of cat and mouse with them.

Approximately one mile south of Indian Head, they said, the object rapidly descended toward their vehicle causing Barney to stop directly in the middle of the highway. The huge, silent craft hovered approximately 80–100 feet (24–30 m) above the Hills' 1957 Chevrolet Bel Air and filled the entire field of the windshield. It reminded Barney of a huge pancake. Carrying his pistol in his pocket, he stepped away from the vehicle and moved closer to the object. Using the binoculars, Barney claimed to have seen about 8 to 11 humanoid figures who were peering out of the craft's windows, seeming to look at him. In unison, all but one figure moved to what appeared to be a panel on the rear wall of the hallway that encircled the front portion of the craft. The one remaining figure continued to look at Barney and communicated a message telling him to "stay where you are and keep looking."

Barney had a conscious, continuous recollection of observing the humanoid forms wearing glossy black uniforms and black caps. Red lights on what appeared to be bat-wing fins began to telescope out of the sides of the craft and a long structure descended from the bottom of the craft. The silent craft approached to what Barney estimated was within 50–80 feet (15–24 m) overhead and 300 feet (91 m) away from him. On October 21, 1961, Barney reported to NICAP Investigator Walter Webb, that the "beings were somehow not human".

Barney "tore" the binoculars away from his eyes and ran back to his car. In a near hysterical state, he told Betty, "They're going to capture us!" He saw the object again shift its location to directly above the vehicle. He drove away at high speed, telling Betty to look for the object. She rolled down the window and looked up, but saw only darkness above them, even though it was a bright, starry night.

Almost immediately, the Hills heard a rhythmic series of beeping or buzzing sounds which they said seemed to bounce off the trunk of their vehicle. The car vibrated and a tingling sensation passed through the Hills' bodies. Betty touched the metal on the passenger door expecting to feel an electric shock, but felt only the vibration. The Hills said that at this point in time they experienced the onset of an altered state of consciousness that left their minds dulled. A second series of codelike beeping or buzzing sounds returned the couple to full consciousness. They found that they had traveled nearly 35 miles (56 km) south but had only vague, spotty memories of this section of road. They recalled making a sudden unplanned turn, encountering a roadblock, and observing a fiery orb in the road.

Arriving home at about dawn, the Hills assert that they had some odd sensations and impulses they could not readily explain: Betty insisted their luggage be kept near the back door rather than in the main part of the house. Their watches would never run again. Barney noted that the leather strap for the binoculars was torn, though he could not recall it tearing. The toes of his best dress shoes were inexplicably scraped. Barney says he was compelled to examine his genitals in the bathroom, though he found nothing unusual. They took long showers to remove possible contamination and each drew a picture of what they had observed. Their drawings were strikingly similar.

Perplexed, the Hills say they tried to reconstruct the chronology of events as they witnessed the UFO and drove home. But immediately after they heard the buzzing sounds, their memories became incomplete and fragmented. They vaguely recalled a luminous moon shape sitting on the road. Barney recalled saying "Oh no, not again". Betty thought Barney had taken a sharp left turn off Route 3.

After sleeping for a few hours, Betty awoke and placed the shoes and clothing she had worn during the drive into her closet, observing that the dress was torn at the hem, zipper and lining. Later, when she retrieved the items from her closet, she noted a pinkish powder on her dress. She hung the dress on her clothesline and the pink powder blew away. But the dress was irreparably damaged. She threw it away, but then changed her mind, retrieving the dress and hanging it in her closet. Over the years, five laboratories have conducted chemical and forensic analyses on the dress.

There were shiny, concentric circles on their car's trunk that had not been there the previous day. Betty and Barney experimented with a compass, noting that when they moved it close to the spots, the needle would whirl rapidly. But when they moved it a few inches away from the shiny spots, it would drop down.

04-24-1964 The **Lonnie Zamora incident** was a UFO close encounter of the third kind which occurred on Friday, April 24, 1964, at about 5:50 p.m., on the southern outskirts of Socorro, New Mexico. Several primary witnesses emerged to report stages and aspects of the event, which included the craft's approach, din, conspicuous flame, and physical evidence left behind immediately afterward. It was however Lonnie Zamora, a New Mexico State police officer who was on duty at the time, who came closest to the object and provided the most prolonged and comprehensive account. Some physical trace evidence left behind—burned vegetation and soil, ground landing impressions, and metal scrapings on a broken rock in one of the impressions—was subsequently observed and analyzed by investigators for the military, law enforcement, and civilian UFO groups.

The event and its body of evidence is sometimes deemed one of the best documented, yet most perplexing UFO reports. It was immediately investigated by the U.S. Army, U.S. Air Force, and FBI, and received considerable coverage in the mass media. It was one of the cases that helped persuade astronomer J. Allen Hynek, one of the primary investigators for the Air Force, that some UFO reports represented an intriguing mystery. After extensive investigation, the AF's Project Blue Book was unable to come up with a conventional explanation and listed the case as an "unknown".

Alone in his patrol car, Sergeant Lonnie Zamora was chasing a speeding car due south of Socorro, New Mexico on April 24, 1964, at about 5:45 p.m., when he "heard a roar and saw a flame in the sky to southwest some distance away — possibly a 1/2 mile or a mile." Thinking a local dynamite shack might have exploded, Zamora broke off the chase and went to investigate.

Though Zamora says he did not pay much attention to the flame, that the sun was "to west and did not help vision", and he was wearing green sunglasses over prescription glasses. In interviews with Air Force investigators for Project Blue Book he goes to some lengths to describe the long, narrow, funnel-shaped "bluish orange" flame. He thought there might be some dust at the bottom, and attributed it to the windy day. The weather was "Clear, sunny sky otherwise — just a few clouds scattered over area."

He describes the noise as "a roar, not a blast. Not like a jet. Changed from high frequency to low frequency and then stopped. Roar lasted possibly 10 seconds" as he approached on a gravel road. "Saw flame about as long as heard the sound. Flame same color as best I can recall. Sound distinctly from high to low until it disappeared."

He explains that his car windows were down. Zamora notes no other possible witnesses except possibly the car in front, which he estimates might have heard the noise but not seen the flame because it would be behind the brow of the hill from their viewpoint.

Zamora struggled to get his car up the steep hill. Successful on the third attempt, he noted no further noise. For the next 10–15 seconds he proceeded west, looking for the shack whose precise location he did not recall. It was then that he noticed a shiny object, "to south about 150 to 200 yards", that at first he took to be an "overturned white car ... up on radiator or on trunk", with two people standing close to it, one of whom seemed to notice him with some surprise and gave a start. The shiny object was "like aluminum — it was whitish against the mesa background, but not chrome", and shaped like a letter "O". Having stopped for a couple of seconds, Zamora approached in his car meaning to help.

Zamora only caught a brief sight of the two people in white coveralls beside the "car". He recalls nothing special about them. "I don't recall noting any particular shape or possibly any hats, or headgear. These persons appeared normal in shape — but possibly they were small adults or large kids."

Zamora drove towards the scene, radioing his dispatcher to say he would be out of his car "checking the car in the arroyo." He stopped his car, got out, and attended to the radio microphone, which he had dropped, then he started to approach the object.

09-05-1956 Cisco Grove Encounter. Hunter Donald Schrum claimed he saw a flying white light in the forest and was pursued by men as well as "robots". Condon Committee investigators were skeptical of Schrum's story because of its inconsistencies.

1965 to 1967 Charles Hall. Charles Hall claimed that while working at Nellis Air Force Base in Nevada he had contact with tall white aliens who often traveled to Las Vegas casinos for entertainment with CIA agents.

03-09-1965 The **Exeter incident** was a highly publicized UFO sighting that occurred on September 3, 1965 approximately 5 miles from Exeter, New Hampshire, in the neighboring community of Kensington. Although several separate sightings had been made by numerous witnesses in the weeks leading up to September 3, the specific incident, eventually to become by far the most famous, involved a local teenager and two police officers. The November/December 2011 edition of *Skeptical Inquirer* offers an explanation of the incident, based on details reported by the eyewitnesses.

07-01-1965 The **Exeter incident** was a highly publicized UFO sighting that occurred on September 3, 1965 approximately 5 miles from Exeter, New Hampshire, in the neighboring community of Kensington. Although several separate sightings had been made by numerous witnesses in the weeks leading up to September 3, the specific incident, eventually to become by far the most famous, involved a local teenager and two police officers. The November/December 2011 edition of *Skeptical Inquirer* offers an explanation of the incident, based on details reported by the eyewitnesses.

At approximately 2 am on September 3, 1965, Norman Muscarello was hitchhiking to his parents' home in Exeter along Highway 150. Muscarello, 18, had recently graduated from high school and was about to leave for service in the U.S. Navy. He had been visiting his girlfriend at her parents' home in nearby Amesbury, Massachusetts; since he did not own a car he would catch a ride to and from his girlfriend's home. However, at that time of night there was little traffic on the highway, and as he walked he noticed 5 flashing red lights in some nearby woods. The lights illuminated the woods and a nearby farmhouse (the farm belonged to the Dining family, who were not at home at the time). The lights soon moved towards him, and Muscarello became terrified and dove into a ditch. The lights moved away and hovered near the Dining farmhouse before going back into the woods. Muscarello ran to the farmhouse, pounded on the door and yelled for help, but no one answered. When he saw a car coming down the road, he ran into the road and forced it to stop. The couple in the car drove him to the Exeter police station.

At the station Muscarello told his story to police officer Reginald Toland, who worked the night desk at the police station. Toland, who knew Muscarello, was impressed by his obvious fear and agitated state. Toland radioed police officer Eugene Bertrand, Jr., who had earlier in the evening passed a distressed woman sitting in her car on Highway 108. When Bertrand stopped and asked if she had a problem, the woman told him that a "huge object with flashing red lights" had been following her car for 12 miles and stopped over her car before flying away. Bertrand considered her a "kook" but did stay with her for approximately 15 minutes until she had calmed down and was ready to resume her drive. After arriving at the police station and hearing Muscarello's story, Bertrand decided to drive back to the Dining farm with Muscarello to investigate the field where he had seen the UFO.

07-01-1965 July 1, 1965, Valensole, Alpes-de-Haute-Provence. An alleged UFO sighting and close encounter by farmer Maurice Masse. According to Masse, he encountered two small beings near a spherical vehicle that had landed in a nearby field. Masse claims that he was paralyzed when one of the beings pointed a tube-like object towards him.
Masse said he watched the beings looking at plants and making grunting
sounds until they returned to the vehicle and flew away. According to
his wife, Masse said he received some kind of communication from the
beings, considered his encounter "a spiritual experience", and looked
upon the site as "hallowed ground" that "should be kept in his family
forever". UFOlogists consider Masse's claims significant and cite "landing gear impressions" found in the soil

09-16-1965 1965 Craft Landing. Police constables Lockem and de Klerk observed a landed, copper-colored craft on the Pretoria–Bronkhorstspruit freeway. It was described as being 30 feet (9.1 m) wide and left a 6 feet (1.8 m) wide imprint on the asphalt after a rapid, fiery ascent. The incident was confirmed in a press release by Lt. Colonel J.B. Brits, district commandant for Pretoria North.

12-09-1965 The **Kecksburg UFO incident** occurred on December 9, 1965, at Kecksburg, Pennsylvania, USA. A large, brilliant fireball was seen by thousands in at least six U.S. states and Ontario, Canada. It streaked over the Detroit, Michigan – Windsor, Ontario area, reportedly dropped hot metal debris over Michigan and northern Ohio, starting some grass fires, and caused sonic booms in the Pittsburgh metropolitan area. It was generally assumed and reported by the press to be a meteor after authorities discounted other proposed explanations such as a plane crash, errant missile test, or reentering satellite debris. However, eyewitnesses in the small village of Kecksburg, about 30 miles southeast of Pittsburgh, claimed something crashed in the woods. A boy said he saw the object land; his mother saw a wisp of blue smoke arising from the woods and alerted authorities. Another reported feeling a vibration and "a thump" about the time the object reportedly landed. Others from Kecksburg, including local volunteer fire department members, reported finding an object in the shape of an acorn and about as large as a Volkswagen Beetle. Writing resembling Egyptian hieroglyphics was also said to be in a band around the base of the object. Witnesses further reported that intense military presence, most notably the United States Army, secured the area, ordered civilians out, and then removed an object on a flatbed truck. The military claimed they searched the woods and found "absolutely nothing."

The *Tribune-Review* from nearby Greensburg which had a reporter at the scene ran an article the next morning, "Unidentified Flying Object Falls near Kecksburg—Army Ropes off Area". The article continued, "The area where the object landed was immediately sealed off on the order of U.S. Army and State Police officials, reportedly in anticipation of a 'close inspection' of whatever may have fallen ...

State Police officials there ordered the area roped off to await the expected arrival of both U.S. Army engineers and possibly, civilian scientists." However, a later edition of the newspaper stated that nothing had been found after authorities searched the area.

The official explanation of the widely seen fireball was that it was a mid-sized meteor. However speculation as to the identity of the Kecksburg object (if there was one—reports vary) range from alien craft to debris from Kosmos 96, a Soviet space probe intended for Venus but which failed and never left the Earth's atmosphere.

Several articles were written about the fireball in science journals. The February 1966 issue of *Sky & Telescope* reported that the fireball was seen over the Detroit-Windsor area at about 4:44 p.m. EST. The Federal Aviation Administration had received 23 reports from aircraft pilots, the first starting at 4:44 p.m. A seismograph 25 miles southwest of Detroit had recorded the shock waves created by the fireball as it passed through the atmosphere. The *Sky & Telescope* article concluded that "the path of the fireball extended roughly from northwest to southeast" and ended "in or near the western part of Lake Erie".

A 1967 article by two astronomers in the *Journal of the Royal Astronomical Society of Canada* (JRASC) used the seismographic record to pinpoint the time of passage over the Detroit area to 4:43 p.m. In addition, they used photographs of the trail taken north of Detroit at two different locations to triangulate the trajectory of the object. They concluded that the fireball was descending at a steep angle, moving from the southwest to the northeast, and likely impacted on the northwestern shore of Lake Erie near Windsor, Ontario.

The JRASC trajectory was at nearly right angles to that proposed by *Sky & Telescope*, or a trajectory that would have taken the fireball in the direction of western Pennsylvania and Kecksburg. Thus, if the calculation was correct, this would rule out the fireball being involved with what may or may not have happened in Kecksburg. The JRASC article is often cited by skeptics to debunk the notion of a UFO crash at Kecksburg. However, the JRASC article has been criticized as lacking any error analysis. Since the triangulation base used by the astronomers in their calculations was very narrow, even very small errors in determination of directions could result in a very different triangulated trajectory. Measurement errors of slightly more than one-half degree would make possible a straight-line trajectory towards the Kecksburg area and a much shallower angle of descent than reported in the JRASC article. It was also pointed out that the photos used actually show the fireball trail becoming progressively thinner, suggesting motion away from the cameras, or in the direction of Pennsylvania. Had the trajectory been sideways to the cameras, as contended in the JRASC article, the trail would likely have remained roughly constant in thickness.

A reporter and news director for the local radio station WHJB, John Murphy, arrived on the scene of the event before authorities had arrived, in response to several calls to the station from alarmed citizens. He took several photographs and conducted interviews with witnesses. His former wife Bonnie Milslagle later reported that all but one roll of the film were confiscated by military personnel. WHJB office manager Mabel Mazza described one of the pictures: "It was very dark and it was with a lot of trees around and everything. And I don't know how far away from the site he was. But I did see a picture of a sort of a cone-like thing. It's the only time I ever saw it."

In the following weeks, Murphy became enveloped with the incident and wrote a radio documentary called *Object in the Woods*, featuring his experiences and interviews he had conducted that night. Shortly before the documentary would have aired, he received an unexpected visit at the station from two men in black suits identifying themselves as government officials. They requested to speak with him in a back room behind closed doors. The meeting lasted about 30 minutes. A WHJB employee, Linda Foschia, recalled the men confiscated some of Murphy's audio tapes from that night, and that no one knows what happened to the remaining photographs. A week after the visit, an agitated Murphy aired a censored version of the documentary, which he claimed in its introduction had to be edited due to some interviewees requesting their statements be removed from the broadcast in fear of getting in trouble with the police and Army. The new version contained nothing revealing, and did not mention an object at all. Mazza, and also Murphy's wife, remember the aired documentary was entirely different from what Murphy had originally written. (See pp. 4–5 of CFI's report in External links for details of the aired documentary.)

After the airing, Murphy became uncharacteristically despondent and completely stopped all investigation on the case and refused to talk to anyone about it again, and never gave clear reasons why. In February 1969, Murphy was struck and killed by an unidentified car in an apparent hit-and-run while crossing a road. The hit-and-run occurred near Ventura, California, while Murphy was on vacation.

There had been some speculation (e.g. NASA's James Oberg) that the object in the Kecksburg Incident may have been debris from Kosmos 96, a Soviet satellite. Kosmos 96 had a bell- or acorn-like shape similar to the object reported by eyewitnesses (though much smaller than witnesses reported).

However, in a 1991 report, US Space Command concluded that Kosmos 96 crashed in Canada at 3:18 a.m. on December 9, 1965, about 13 hours before the fireball thought to be the Kecksburg object undergoing re-entry was recorded at 4:45 p.m.

In addition, in a 2003 interview Chief Scientist for Orbital Debris at the NASA Johnson Space Center Nicholas L. Johnson stated:

I can tell you categorically, that there is no way that any debris from Kosmos 96 could have landed in Pennsylvania anywhere around 4:45 p.m. [...] That's an absolute. Orbital mechanics is very strict.

Michael Ryan

11-02-1966 **Mothman** is a humanoid moth-like creature reportedly seen in the Point Pleasant area of West Virginia from November 15, 1966, to December 15, 1967. The first newspaper report was published in the *Point Pleasant Register* dated November 16, 1966, titled "Couples See Man-Sized Bird...Creature...Something". The being subsequently entered regional folklore.

Mothman was introduced to a wider audience by Gray Barker in 1970, later popularized by John Keel in his 1975 book *The Mothman Prophecies*, claiming that Mothman was related to a wide array of supernatural events in the area and the collapse of the Silver Bridge. The 2002 film *The Mothman Prophecies*, starring Richard Gere, was based on Keel's book.

On November 12, 1966, five men who were digging a grave at a cemetery near Clendenin, WV, claimed to see a man-like figure fly low from the trees over their heads. This is often identified as the first known sighting of what became known as the Mothman.

Shortly thereafter, on November 15, 1966, two young couples from Point Pleasant, Roger and Linda Scarberry, and Steve and Mary Mallette told police they saw a large white creature whose eyes "glowed red" when the car headlights picked it up. They described it as a "large flying man with ten-foot wings following their car while they were driving in an area outside of town known as 'the TNT area', the site of a former World War II munitions plant.

During the next few days, other people reported similar sightings. Two volunteer firemen who sighted it said it was a "large bird with red eyes". Mason County Sheriff George Johnson commented that he believed the sightings were due to an unusually large heron he termed a "shitepoke". Contractor Newell Partridge told Johnson that when he aimed a flashlight at a creature in a nearby field its eyes glowed "like bicycle reflectors", and blamed buzzing noises from his television set and the disappearance of his German Shepherd dog on the creature. Wildlife biologist Dr. Robert L. Smith at West Virginia University told reporters that descriptions and sightings all fit the sandhill crane, a large American crane almost as high as a man with a seven-foot wingspan featuring circles of reddish coloring around the eyes, and that the bird may have wandered out of its migration route.

There were no Mothman reports in the immediate aftermath of the December 15, 1967, collapse of the Silver Bridge and the death of 46 people, giving rise to legends that the Mothman sightings and the bridge collapse were connected.

Folklorist Jan Harold Brunvand notes that Mothman has been widely covered in the popular press, some claiming sightings connected with UFOs, and others claiming that a military storage site was Mothman's "home". Brunvand notes that recountings of the 1966-67 Mothman reports usually state that at least 100 people saw Mothman with many more "afraid to report their sightings" but observed that written sources for such stories consisted of children's books or sensationalized or undocumented accounts that fail to quote identifiable persons. Brunvand found elements in common among many Mothman reports and much older folk tales, suggesting that something real may have triggered the scares and became woven with existing folklore.

He also records anecdotal tales of Mothman supposedly attacking the roofs of parked cars inhabited by teenagers.

Some ufologists, paranormal authors, and cryptozoologists claim that Mothman was an alien, a supernatural manifestation, or an unknown cryptid. In his 1975 book *The Mothman Prophecies*, author John Keel claimed that the Point Pleasant residents experienced precognitions including premonitions of the collapse of the Silver Bridge, unidentified flying object sightings, visits from inhuman or threatening men in black, and other phenomena.

Skeptic Joe Nickell says that a number of hoaxes followed the publicity generated by the original reports, such as a group of construction workers who tied flashlights to helium balloons. Nickell attributes the Mothman reports to pranks, misidentified planes, and sightings of a barred owl, an albino owl, suggesting that the Mothman's "glowing eyes" were actually red-eye effect caused from the reflection of light from flashlights or other bright light sources. The area lies outside the snowy owl's usual range.

03-1966 Michigan Swamp Gas Sightings. Widely reported wave of sightings attributed to swamp gas by J. Allen Hynek.

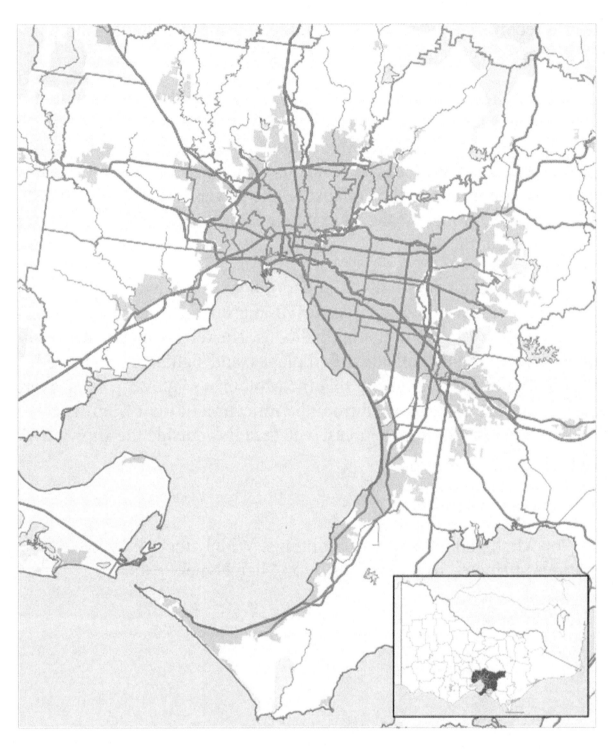

04-06-1966 The **Westall UFO encounter** is an event that occurred on 6 April 1966 in Melbourne, Victoria, Australia. Around 11.00 am, for about 20 minutes, more than 200 students and teachers at two Victorian state schools allegedly witnessed an unexplained flying object which descended into a nearby open wild grass field. The paddock was adjacent to a grove of pine trees in an area known as The Grange (now a nature reserve). According to reports, the object then ascended in a north-westerly direction over the suburb of Clayton South, Victoria, Australia.

At approximately 11.00 am on Wednesday, 6 April 1966, a class of students and a teacher from Westall High School (now Westall Secondary College) were just completing a sport activity on the main oval when an object, described as being a grey saucer-shaped craft with a slight purple hue and being about twice the size of a family car, was alleged to have been seen. Witness descriptions were mixed: Andrew Greenwood, a science teacher, told *The Dandenong Journal* at the time that he saw a silvery-green disc. According to witnesses the object was descending and then crossed and overflew the high school's south-west corner, going in a south-easterly direction, before disappearing from sight as it descended behind a stand of trees and into a paddock at The Grange in front of the Westall State School (primary students). After a short period (approximately 20 minutes) the object - with witnesses now numbering over 200 - then climbed at speed and departed towards the north-west. As the object gained altitude some accounts describe it as having been pursued from the scene by five unidentified aircraft which circled the object.

04-15-1966 Catalina Island UFO. Cameraman Leman Hanson shot 10 seconds of 16 mm color footage of a UFO.

04-17-1966 Portage County UFO chase. Several police officers pursue what they believe to be a UFO for 30 minutes.

01-25-1967 Betty Andreasson Abduction. A woman claims to have been taken aboard UFOs by aliens many times over a number of years.

05-20-1967 Falcon Lake incident. A man is reported to have been burned by the exhaust of a landed cigar-shaped object.

The **Close encounter of Cussac** is the name given to claims of a close encounter with alien beings by a young brother and sister in Cussac, Cantal, France.

On Aug. 29, 1967, a 13-year-old boy and his 9-year-old sister told local police they were watching cows in a field and saw "four small black beings about 47 inches tall" who appeared to rise in the air and enter "a round spaceship, about 15 feet in diameter" that was hovering over the field. The police noted "sulfur odor and the dried grass" at the place where the sphere was alleged to have taken off. The children's story is one of the reports of UFO sightings investigated by the French government made public in a mass release of documents in March 2007 which received so many hits on its first day that the site crashed.

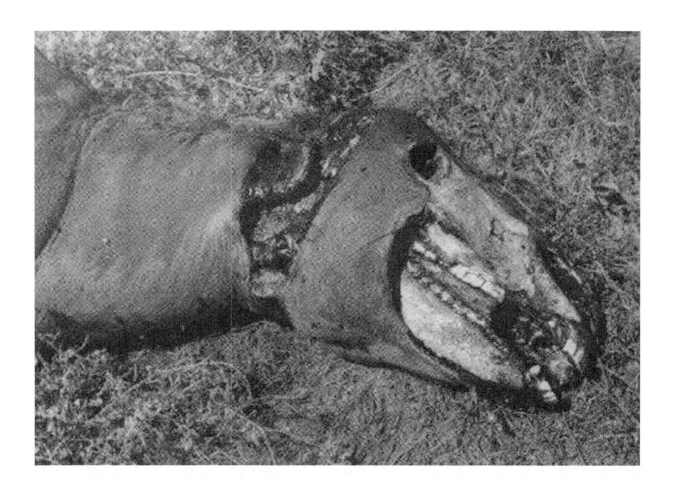

09-07-1967 Snippy the Horse Mutilation. Widely considered to be the first unusual animal death to be related by its witnesses to UFOs and aliens.

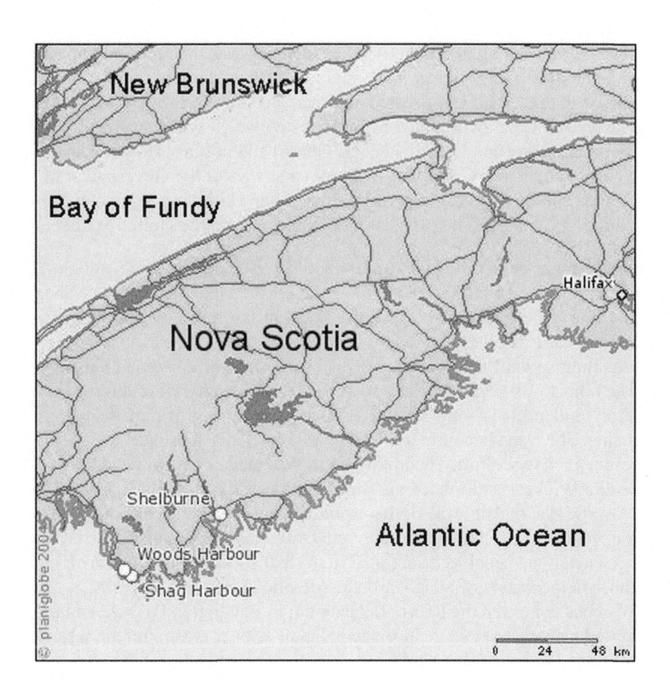

10-04-1967 The **Shag Harbour UFO incident** was the reported impact of an unknown large object into waters near Shag Harbour, a tiny fishing village in the Canadian province of Nova Scotia on October 4, 1967. The reports were investigated by various civilian (Royal Canadian Mounted Police and Canadian Coast Guard) and military (Royal Canadian Navy and Royal Canadian Air Force) agencies of the Government of Canada and the U.S. Condon Committee. Shag Harbour is equivalent in stature in Canada as the Roswell UFO incident is in the United States.

On the night of October 4, 1967, at about 11:20 p.m. Atlantic Daylight Time, it was reported that something had crashed into the waters of Shag Harbour Nova Scotia . At least eleven people saw a low-flying lit object head towards the harbour. Multiple witnesses reported hearing a whistling sound "like a bomb," then a "whoosh," and finally a loud bang. The object was never officially identified, and was therefore referred to as an unidentified flying object (UFO) in Government of Canada documents. The Canadian military became involved in a subsequent rescue/recovery effort. The initial report was made by local resident Laurie Wickens and four of his friends. Driving through Shag Harbour Nova Scotia, on Highway 3, they spotted a large object descending into the waters off the harbour. Attaining a better vantage point, Wickens and his friends saw an object floating 250 m (820 ft) to 300 m (980 ft) offshore in the waters of Shag Harbour, Atlantic Ocean Nova Scotia . Wickens contacted the RCMP detachment in Barrington Passage and reported he had seen a large airplane or small airliner crash into the waters off Shag Harbour Nova Scotia

Assuming an aircraft had crashed, within about 15 minutes, 10 RCMP officers arrived at the scene. Concerned for survivors, the RCMP detachment contacted the Rescue Coordination Centre (RCC) in Halifax to advise them of the situation, and ask if any aircraft were missing. Before any attempt at rescue could be made, the object started to sink and disappeared from view. A rescue mission was quickly assembled. Within half an hour of the crash, local fishing boats went out to the crash site in the waters of the Gulf of Maine off Shag Harbour to look for survivors. No survivors, bodies or debris were taken, either by the fishermen or by a Canadian Coast Guard search and rescue cutter, which arrived about an hour later from nearby Clark's Harbour. By the next morning, RCC Halifax had determined that no aircraft were missing. While still tasked with the search, the captain of the Canadian Coast Guard cutter received a radio message from RCC Halifax that all commercial, private and military aircraft were accounted for along the eastern seaboard, in both Atlantic Canada and New England. The same morning, RCC Halifax also sent a priority telex to the "Air Desk" at Royal Canadian Air Force headquarters in Ottawa, which handled all civilian and military UFO sightings, informing them of the crash and that all conventional explanations such as aircraft, flares, etc. had been dismissed. Therefore, this was labeled a "UFO Report." The head of the Air Desk then sent another priority telex to the Royal Canadian Navy headquarters concerning the "UFO Report" and recommended an underwater search be mounted. The RCN in turn sent another priority telex tasking Fleet Diving Unit Atlantic with carrying out the search.

Two days after the incident had been observed, a detachment of RCN divers from Fleet Diving Unit Atlantic was assembled and for the next three days they combed the seafloor of the Gulf of Maine off Shag Harbor looking for an object. The final report said no trace of an object was found.

02-18-1967 Vashon Island incident. Three men reported seeing a shiny saucer shaped craft, which froze a pond over which it had hovered.

08-07-1967 Buff Ledge Camp Abduction. Two teenage employees of a summer camp reported sighting UFOs over the lake and claimed to have experienced missing time.

11-28-1968 The Sverdlovsk Midget. Near Sverdlovsk discovered a falling object in the Roshevski forest. Some claimed that about the UFO and a Soviet Alien-autopsy survived some film reel. About the incident made a documentary film with Roger Moore.

The **Jimmy Carter UFO Incident** is the name given to an incident in which Jimmy Carter (US President 1977-1981) reported seeing an unidentified flying object while at Leary, Georgia, in 1969.

While governor of Georgia, Carter was asked to file a report of the sighting by the International UFO Bureau in Oklahoma City, Oklahoma, which he did in September 1973. Since its writing, the report has been discussed several times by both ufologists and by members of the mainstream media. Carter does not think that it was an alien spacecraft.

One evening in 1969, two years before he became governor of Georgia, Carter was preparing to give a speech at a Lions Club meeting. At about 7:15 p.m (EST), one of the guests called his attention to a strange object that was visible about 30 degrees above the horizon to the west of where he was standing. Carter described the object as being bright white and as being about as bright as the moon. It was said to have appeared to have closed in on where he was standing but to have stopped beyond a stand of pine trees some distance from him. The object is then said to have changed color, first to blue, then to red, then back to white, before appearing to recede into the distance.

Carter felt that the object was self-luminous, but not a solid in nature. Carter's report indicates that it was witnessed by about ten or twelve other people, and was in view for ten to twelve minutes before it passed out of sight.

In 1973 Carter said (Sheaffer 1998:20–21)

"There were about twenty of us standing outside of a little restaurant, I believe, a high school lunch room, and a kind of green light appeared in the western sky. This was right after sundown. It got brighter and brighter. And then it eventually disappeared. It didn't have any solid substance to it, it was just a very peculiar-looking light. None of us could understand what it was."

Speaking in a 2005 interview, Carter states:

"all of a sudden, one of the men looked up and said, 'Look, over in the west!' And there was a bright light in the sky. We all saw it. And then the light, it got closer and closer to us. And then it stopped, I don't know how far away, but it stopped beyond the pine trees.

148

And all of a sudden it changed color to blue, and then it changed to red, then back to white. And we were trying to figure out what in the world it could be, and then it receded into the distance."

The exact date on which the sighting occurred has been called into question by investigators. According to the report that he filed with the International UFO Bureau four years after the incident, Carter saw the UFO in October 1969. However investigators have cited Lions Club records as evidence that it occurred nine months earlier.

According to a meeting report that he filed with the Lions Club, Carter gave his Leary speech on January 6, 1969, not in October. The setting of his January meeting as described in his report to the Lions Club also matches the setting that he would later describe to the media when speaking about his sighting. His report to the Lions Club made no mention of the sighting itself.

Other evidence uncovered rules out the October 1969 date and is consistent with January 1969. First, Carter visited the Leary Lions Club in his capacity as district governor of the Lions Club. His term ended in June 1969. Second, the Leary Lions Club disbanded several months before October 1969

According to an investigation carried out in 1976, some seven years after the event, most of those present at the meeting either did not recall the event, or did not recall it as being anything important. According to Fred Hart, the only guest contacted who remembered seeing the object: "It seems like there was a little—like a blue light or something or other in the sky that night—like some kind of weather balloon they send out or something ... it had been pretty far back in my mind."

While puzzled by the object and its origins Carter, himself, later said that while he had considered the object to be a UFO—on the grounds it was unexplained—his knowledge of physics had meant he had not believed himself to be witnessing an alien spacecraft.

On January 6, 1969 the sky was clear in Leary and the planet Venus was near its maximum brightness and in the direction described by Carter. Ufologist Robert Sheaffer concluded that the object that Carter witnessed was a misidentification of Venus. Ufologist Allan Hendry did calculations and agrees with the assessment of it being Venus. This could also be the Venus "Halo", as was discussed on *The Skeptics' Guide to the Universe* podcast #105 in a 2007 interview with Jimmy Carter. In the interview Carter stated that he did not believe the object was Venus, explaining that he was an amateur astronomer and knew what Venus looked like. He also said that as a scientist he did not believe it was an alien craft and at the time assumed it was probably a military aircraft from a nearby base. However, he said that the object did not make any sound like a helicopter would do. Carter also said that he did not believe that any extraterrestrials have visited Earth. He also stated he knows of no government cover-up of extraterrestrial visits and that the rumors that the CIA refused to give him information about UFOs are not true.

01-01-1970 Cowichan District Hospital UFO. Nurse Doreen Kendall claims she looked out a hospital window and saw a UFO containing occupants seated at a control panel. Another nurse called over by Kendall saw only a bright, featureless light.

06-27-1972 Bennie Smith, the owner of a farm near Fort Beaufort in the eastern Cape region of South Africa, says he fired shots at an unknown object hovering at treetop height after a worker named Boer de Klerk alerted him to it. Smith believed his shots were accurately aimed, but had no effect. Police sergeant Piet Kitching and police station commander Van Rensburg stated they arrived and fired shots at the object, described as metallic and shaped like a 44-gallon drum with three legs that changed colors before it flew away. They said they found imprints and markings on the ground they believed were made by the object. It is claimed that the Grahamstown army regiment investigated the site, but the base has no records of such an event. The incident received coverage by international press, and led to businesses capitalizing on the incident, with a tavern calling itself the "UFO Bar" and painting flying saucers on the walls and the local Savoy Hotel keeping clippings of the stories posted on its walls. In a humorous editorial, the *New Scientist* stated the apartheid South African government was "very fastidious about the sort of immigrants she welcomes and little green men may very well be on the prohibited list"

11-12-1972 a resident Indian woman reported a disc-shaped object taking off from the Groendal Nature Reserve near Uitenhage, eastern Cape. Three days later, on 2 October, four school boys from Despatch, aged 12 to 16, observed three silvery-clad men in the reserve while hiking. Two of the men arrived from the direction of a shining object, and joined a third to ascend a steep incline on what seemed to be fins, before all disappeared. A set of 9 regular imprints found a month later, was deemed related to the reflective object.

05-1973 The Judy Doraty Abduction. While under hypnosis, Judy Doraty claimed to have been abducted by aliens and to have witnessed a cattle mutilation.

09-20-1973 Skylab 3 UFO Encounter. Blurred photographs of an unknown object were taken by astronauts.

10-1973 Jeff Greenhaw. Alabama policeman Jeff Greenhaw claims he encountered a creature dressed in a shiny metal costume while investigating a UFO report.

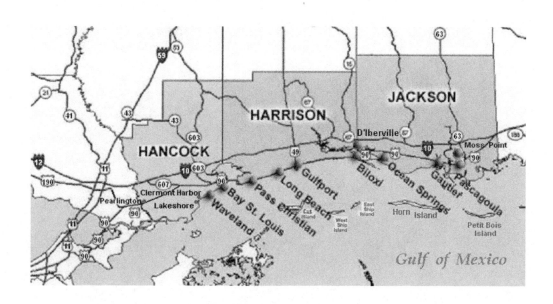

The **Pascagoula Abduction** is a purported UFO sighting and alien abduction alleged to have occurred in 1973 when co-workers Charles Hickson and Calvin Parker claimed they were abducted by aliens while fishing near Pascagoula, Mississippi. The incident earned substantial mass media attention.

On the evening of October 11, 1973, co-workers 42-year-old Charles Hickson and 19-year-old Calvin Parker told the Jackson County, Mississippi Sheriff's office they were fishing off a pier on the west bank of the Pascagoula River in Mississippi when they heard a whirring/whizzing sound, saw two flashing blue lights and an oval shaped object 30-40 feet across and 8-10 feet high. Parker and Hickson claimed that they were "conscious but paralyzed" while three "creatures" took them aboard the object and subjected them to an examination before releasing them.

Charlie Hickson   Calvin Parker

10-11-1973 The **Pascagoula Abduction** is a purported UFO sighting and alien abduction alleged to have occurred in 1973 when co-workers Charles Hickson and Calvin Parker claimed they were abducted by aliens while fishing near Pascagoula, Mississippi. The incident earned substantial mass media attention.On the evening of October 11, 1973, co-workers 42-year-old Charles Hickson and 19-year-old Calvin Parker told the Jackson County, Mississippi Sheriff's office they were fishing off a pier on the west bank of the Pascagoula River in Mississippi when they heard a whirring/whizzing sound, saw two flashing blue lights and an oval shaped object 30-40 feet across and 8-10 feet high. Parker and Hickson claimed that they were "conscious but paralyzed" while three "creatures" took them aboard the object and subjected them to an examination before releasing them.

10-17-1973 Eglin Air Force Base Sighting. An unidentified object was tracked by a Duke Field radar unit during the same time period, and within the same area, that 10 to 15 people observed four strange objects flying in formation between Milton, Florida, and Crestview, Florida, along Interstate 10, according to Eglin officials. Reports from the base indicated that a bright glowing ball of light could be seen travelling parallel with an Air Force C-130 aircraft but at a much higher altitude.

01-23-1974 on the Berwyn Mountains in Llandrillo, Merionethshire, North Wales, lights and noises were observed that were alleged to be related to a UFO sighting on Cadair Berwyn and Cadair Bronwen. Scientific evidence indicates that the event was generated by an earthquake combined with sightings of a bright meteor widely observed over Wales and Northern England at the time.

05-31-1974 Abduction Event. travelers Peter and Frances MacNorman claimed an abduction event on 31 May, which would have started near Fort Victoria in the current southern Zimbabwe, and which would have continued to the vicinity of Beitbridge on the South African border.

08-25-1974 Coyame UFO incident. An alleged mid-air collision between a UFO and a small airplane, near the town of Coyame, Chihuahua, close to the U.S.-Mexico border. Some people believe that a UFO crashed and was secretly recovered, but according to local residents, UFO rumors arose after Mexican military authorities conducted a search to recover drugs and money scattered across the landscape in the wreckage of a plane piloted by a U.S. resident in 1980.

10-30-1975 **Wurtsmith Air Force Base** is a decommissioned United States Air Force base in northeastern Iosco County in the U.S. state of Michigan. The former base includes 4,626 acres (1,872 ha) located approximately two miles west of Lake Huron in the Charter Township of Oscoda, bordered by Van Ettan Lake, the Au Sable State Forest.

NATL. ENQUIRER   DEC. 16 '75

5 Witnesses Pass Lie Test While Claiming . . .

Arizona Man Captured by UFO

In one of the most baffling cases ever recorded, a young laborer was taken aboard a UFO — in full view of six terrified co-workers — and held for five days.

The stunned witnesses readily agreed to take lie-detector tests. Five passed, proving they were telling the truth. The sixth was so nervous that the results of his test were inconclusive.

The abducted man, 22-year-old Travis Walton of Snowflake, Ariz., suddenly disappeared when struck by a dazzling ray from the strange hovering, saucer-shaped object, his fellow workers told police and The ENQUIRER.

In gripping detail, the witnesses described the chilling incident:

Dwayne Smith, 21: "It was a spaceship, there's no doubt of that — and Travis went on it. He got out of our truck, walked toward it — and just vanished!"

Kenneth Peterson, 25: "I saw a bluish light come from the machine and Travis went flying — like he'd touched a hot wire."

Alan Dalis, 21: "It sent out a blue ray, and the last we saw of Travis was his silhouette outlined, arms outstretched. We couldn't believe what was happening — the horror was unreal!"

Mike Rogers, 28: "We were all scared! We couldn't think of any other explanation except that Travis had been taken off by the saucer."

John Goulette, 21: "I know

By TONY BRENNA, JOHN M. CATHCART, CHRIS FULLER, PAUL JENKINS, NICK LONGHURST, ROBERT G. SMITH and JEFF WELLS

SKETCH of hovering UFO drawn by Mike Rogers and Dwayne Smith.

tree trimmers were heading home at dusk along an isolated mountain road. Crew member Dwayne Smith said they were about 12 miles from Heber when they suddenly spotted the saucer hovering in a clearing beside the road.

"I was numb with disbelief — and terrified!" said Smith. "The UFO was smooth, and was giving off a yellowish-orange light.

"We watched it in shock a few seconds, then suddenly just vanished! Mike Rogers, who was driving the truck, screamed 'Shut the door!' and gunned past the saucer.

"When we could see the saucer wasn't following us, Mike stopped the truck and we all got out, shouting and screaming at each other with fear in our faces and terror in our hearts. Then we saw a flash in the trees, and figured the saucer was leaving.

"We went back to the spot where the saucer had been . . . but Travis was gone. He went on the spaceship, there's no doubt of that.

"We went and reported what had happened. We didn't expect anyone to believe us, and nobody did — until we took lie detector tests."

Because Walton wasn't able to recall all the details of his amazing experience, The ENQUIRER arranged for him to be interviewed under hypnosis by Dr. Harder, a member of The ENQUIRER's prestigious

Dr. Harder that 'as he approached the craft for a closer look, he was hit by something and suddenly everything went black.'

'When I woke up, there was a strong light in my eyes and I had problems focusing. I was panicked because there was a terrible pain in my head and chest," he told Dr. Harder.

"My mind cleared a little and I thought I was in a hospital. I was lying on a table on my back, and these figures were standing over me.

"It was weird. They weren't human — they were creatures.

"They looked like well-developed fetuses to me — they were about 5 feet tall and wore tight-fitting tan-brown robes. Their skin was white like a mushroom and they had no clear features. They made no sounds.

"Their faces had no texture or color, and there was no hair. Their foreheads were domed and their eyes were very large. They had long fingers — but no fingernails.

"I panicked and jumped up, knocking a clear plastic tray

EXPERTS QUIZ abducted man, Travis Walton (center). At left is Dr. Jean Rosenbaum, at right Dr. James Harder.

SHERIFF Marlin Gillespie: "I'm sure they saw UFO."

ing aboard the craft — and his description of the creatures he saw fitted the description given by Walton.

"There's been no publicity about this case because the man feels it would jeopardize his job to discuss it."

11-05-1975 **Travis Walton** (born February 10, 1953) is an American logger who was allegedly abducted by a UFO on November 5, 1975, while working with a logging crew in the Apache-Sitgreaves National Forest in Arizona. Walton could not be found, but reappeared after a five-day search.

The Walton case received mainstream publicity and remains one of the best-known instances of alleged alien abduction. UFO historian Jerome Clark writes that "Few abduction reports have generated as much controversy" as the Walton case. It is furthermore one of the very few alleged alien abduction cases with some corroborative eyewitnesses, and one of few alleged abduction cases where the time allegedly spent in the custody of aliens plays a rather minor role in the overall account.

UFO researchers Jenny Randles and Peter Houghe write that "Neither before or since has an abduction story begun in the manner related by Walton and his coworkers. Furthermore, the Walton case is singular in that the victim vanished for days on end with police squads out searching … it is an atypical 'Close Encounter: Fourth Kind' (CE4) … which bucks the trend so much that it worried some investigators; others defend it staunchly."

The case began on Wednesday, November 5, 1975. Then 22 years old, Walton was employed by Mike Rogers, who had for nine years contracted with the United States Forest Service for various duties. Rogers and Walton were best friends; Walton dated Rogers' sister Dana, whom he later married. Others on the crew were Ken Peterson, John Goulette, Steve Pierce, Allen Dallis and Dwayne Smith. They all lived in the town of Snowflake, Arizona.

Rogers was hired to thin out scrub brush and undergrowth from a large area (more than 1,200 acres) near Turkey Springs, Arizona. The job was the most lucrative contract Rogers had received from the Forest Service, but the job was behind schedule. As a result, they worked overtime to fulfill the contract, typically from 6 a.m. until sunset.

Just after 6 p.m. on November 5, Rogers and his crew finished their work for the day and piled into Rogers' truck for the drive back to Snowflake. The crew reported that shortly after beginning the drive home, they saw a bright yellowish light from behind a hill. They drove closer and said they saw a large golden disc hovering above a clearing and shining brightly. It hovered below the tops of the trees about 15 feet (4.6 m) over a pile of logging slash. It was around 8 feet (2.4 m) high and 20 feet (6.1 m) in diameter.

Rogers stopped the truck and Walton leaped out and ran toward the disc. The others said they shouted at Walton to come back but he continued toward the disc. They noticed Walton stepping backwards. The men in the truck reported that Walton was nearly below the object when the disc began making noises similar to a loud turbine. The disc then began to wobble from side to side, and Walton began to cautiously walk away from the object.

Jerome Clark wrote that just after Walton moved away from the disc, the others insisted they saw a beam of blue-green light coming from the disc and "strike" Walton. Clark went on to write that Walton "rose a foot into the air, his arms and legs outstretched, and shot back stiffly some 10 feet (3.0 m), all the while caught in the glow of the light. His right shoulder hit the earth, and his body sprawled limply over the ground.

About 7:30 p.m., Peterson called police from Heber, Arizona, near Snowflake. Deputy Sheriff Chuck Ellison answered the telephone; Peterson initially reported only that one of a logging crew was missing. Ellison then met the crew at a shopping center. They related the tale to him — all the men distraught, two of them in tears — and though he was somewhat skeptical of the fantastic account, Ellison would later reflect "that if they were acting, they were awfully good at it."

Ellison notified his superior — Sheriff Marlin Gillespie — who told Ellison to keep the crew in Heber until he could arrive with Officer Ken Coplan to interview the men. In less than an hour, Gillespie and Coplan arrived, and they heard the tale from the crew. Rogers insisted on returning to the scene immediately to search for Walton, with tracking dogs, if possible. No dogs were available, but the police and some of the crew returned to the scene.

Crew members Smith, Pierce and Goulette were too upset to be of much help in a search, so they elected to return to Snowflake and relate the bad news to friends and family.

At the scene, the law enforcement officers became suspicious of the story related by the crew, mainly because there was nothing in the way of physical evidence to back up the account. Though more police and volunteers arrived to search the area, they found not a trace of Walton. Winter nights could be bitterly cold in the mountains, and Walton had worn only jeans, a denim jacket and a shirt; police were worried that Walton could fall victim to hypothermia if he were lost.

Rogers and Sheriff Coplan went to tell the news to Walton's mother, Mary Walton Kellett, who lived on a small ranch at Bear Creek, some 10 miles (16 km) from Snowflake. Rogers told her what had happened, and she asked him to repeat the account. She then asked calmly if anyone other than the police and the eyewitnesses had heard the story. Coplan thought her reserved response was odd; this factor contributed to the growing suspicion among police that something other than a UFO was responsible for Walton's absence. On the other hand, Clark noted that Kellett was known as being generally guarded, and had furthermore raised six children largely by herself under often trying circumstances, which "had long since taught her to not to fly to pieces in the face of crises and tragedies. Yet in the days ahead, as events overwhelmed her, she would show emotion before friends, acquaintances and strangers alike — a fact that would go unmentioned in debunking treatments of the Walton episode."

About 3 a.m., Kellett telephoned Duane Walton, her second-oldest child. He left his home in Glendale, Arizona, and drove to Snowflake.

By morning on November 6, officials and volunteers had scoured the area around the scene where Walton went missing. No trace of him was discovered, and police suspicions were growing that the UFO tale was concocted to cover up an accident or homicide. Saturday morning, Rogers and Duane Walton arrived at Sheriff Gillespie's office "explosively angry" because they had returned to the scene and found no police there. By that afternoon, police were searching for Walton with helicopters, horse-mounted officers, and jeeps.

By Saturday, word of Walton's disappearance had spread internationally. News reporters, ufologists and the curious began travelling to Snowflake. Among the visitors was Fred Sylvanus, a Phoenix UFO investigator, who interviewed Rogers and Duane Walton on Saturday, November 8. While repeatedly expressing worry for Walton's well-being (and criticizing what they saw as a halfhearted search effort by police), both men would make statements that would return to haunt them, when seized upon by critics.

On the recordings made by Sylvanus, Rogers noted that because of Walton's disappearance and the subsequent search, he would be unable to complete his contract with the Forest Service, and he hoped the search for his missing friend would mitigate the situation. Duane Walton reported he and Travis were quite interested in UFOs, and that some twelve years earlier, Duane had witnessed a UFO similar to the one witnessed by the logging crew. Duane reported that he and Travis had both decided that if they had a chance, they would get as close as possible to any UFO they might see. Duane also suggested that Walton would not be injured by the aliens, because "they don't harm people". Without intending to do so, Rogers and Duane Walton had laid "the foundations for an alternative interpretation of the case" with their statements.

Travis would later report that he never had a "keen" interest in UFOs, even after his supposed abduction, but the tape recorded statement of his brother Duane, while Travis was still missing, runs contrary to Travis's statements.

Shortly after the Sylvanus interview, Snowflake town marshal Sanford Flake announced that the entire affair was a prank engineered by Duane and Travis. They had fooled the logging crew by lighting a balloon and "releasing it at the appropriate time". Flake's wife disagreed, suggesting that her husband's story was "just as farfetched as Duane Walton's".

In the meantime, police officers were making repeated visits to Kellett's home; Duane once returned there to find her in tears as she was being questioned in her living room. Duane told the police to leave unless they had something new to relate, or to ask. Duane suggested that she speak with police only on the front porch, which would allow her to end the interview anytime she chose by simply going inside. She did exactly that after Marshal Flake arrived to relate a message, which Clark notes, contributed to the feeling among skeptics that Kellett was "hiding something. Or someone".

Duane also spoke with William H. Spaulding of *Ground Saucer Watch*. Spaulding suggested that if Walton ever returned, GSW could provide a doctor to examine him in confidence.

On Monday, November 10, all of Rogers' remaining crew took polygraph examinations administered by Cy Gilson, an Arizona Department of Public Safety employee. His questions asked if any of the men caused harm to Walton (or knew who had caused Walton harm), if they knew where Walton's body was buried, and if they told the truth about seeing a UFO.

The men all denied harming Walton (or knowing who had harmed him), denied knowing where his body was, and insisted they had indeed seen a UFO.

Excepting Dallis (who had not completed his exam, thus rendering it invalid), Gilson concluded that all the men were truthful, and the exam results were conclusive. Clark quotes from Gilson's official report: "These polygraph examinations prove that these five men did see some object they believed to be a UFO, and that Travis Walton was not injured or murdered by any of these men on that Wednesday". If the UFO was fake, Gilson thought, "five of these men had no prior knowledge of it".

Following the polygraph tests, Sheriff Gillespie announced that he accepted the UFO story, saying "There's no doubt they're telling the truth."

In 2009, Walton was a participant on game show *The Moment of Truth*. When asked if he was abducted by a UFO in 1975, he responded, "Yes", an answer which the polygraph examiner determined to be deceptive prior to taping. Walton, in response to this outcome, said that polygraphs are 97% accurate, even in the best of cases.

Duane remembered Spaulding's promise of a confidential medical examination. Without having notified authorities of Walton's return, Duane drove him to Phoenix, Arizona, late Tuesday morning, where they were to meet with Dr. Lester Steward.

The Waltons reported that they were disappointed to learn that Steward was not a medical doctor as Spaulding had promised, but a hypnotherapist.

Spaulding and Steward would later report that the Waltons had stayed with them for over two hours, while the Waltons insist they were at Steward's office for, at most, 45 minutes, most of which was occupied with trying to determine the nature of Steward's qualifications. The precise time spent with Steward would later become an issue in the case.

By Tuesday afternoon, word of Walton's return had leaked out to the public. Duane took a telephone call from Spaulding, and told Spaulding not to bother the family again. Clark writes that after this telephone call, "Spaulding became a sworn enemy in the case."

Among the other telephone calls after news of Walton's return was one from Coral Lorenzen of Aerial Phenomena Research Organization (APRO), a civilian UFO research group. She promised Duane that she could arrange an examination for Walton by two medical doctors — general practitioner Joseph Saults and pediatrician Howard Kandell — at Duane's home. Duane agreed, and the exam began at about 3:30 p.m. Tuesday.

Clark writes that "between Lorenzen's call and the physicians' examination, another party would enter, and hugely complicate, the story". Lorenzen was telephoned by an employee of the *National Enquirer*, an American tabloid newspaper known for its sensationalistic tone. The *Enquirer* employee promised to finance APRO's investigation, in exchange for APRO's "cooperation and access to the Waltons". Since the Enquirer's financial resources were far greater than APRO's, Lorenzen agreed to the arrangement.

In his survey of UFO abduction literature, Terry Matheson writes that "Walton's experience stands out by virtue of its *not* being particularly bizarre as far as abduction accounts go."

Walton reported that after approaching the UFO near the work site, he heard the spacecraft make a low rumbling sound that sent a powerful wave of vibrations throughout the entire area. The last thing he remembered was being struck by a bright, blinding beam of light. When he woke, Walton said he was in a very small, cramped room on a flat reclined bed, like an operating table. Another bright light shone above him, and the air was heavy and humid. He was also in a lot of pain, and had some trouble breathing, but his first thought was that he was in a normal hospital.

As his faculties returned, Walton says when he came to his senses, he realized he was surrounded by three strange figures that he immediately knew upon observing them, were not human, but humanoid creatures. Each was wearing an orangish brown jumpsuit of a soft material with no seams or buttons. Walton described the beings as the typical so-called Greys which feature in many abduction accounts: "shorter than five feet, and they had completely smooth, bald heads, with no body hair, and their hands had no fingernails. Their heads were domed, very large and disproportionate. They looked like fetuses. They had large eyes —enormous eyes— almost all dark brown, without much white in them. The creepiest thing about them were those eyes, they just stared through me." Their ears, noses and mouths "seemed real small, maybe just because their eyes were so huge."

Walton related that he feared for his safety and got to his feet. Terrified, he shouted at the creatures to stay away. As they moved towards him he grabbed a glass-like cylinder object from a nearby shelf and tried to break its tip to create a makeshift knife, but found the object unbreakable, so instead waved it at the creatures as a weapon. The trio of creatures, seeing Walton had become violent, quickly fled and left him alone in the room.

Matheson finds this portion of the narrative troublingly inconsistent, noting that "despite his 'weakened' condition, 'aching body' and 'splitting pain in his skull', maladies or which no cause is suggested, he has no trouble jumping up from his operating table, seizing a conveniently placed glass-like rod, and, assuming a karate 'fighting stance', frightened them with this display of macho aggression, enough at least to cause them to run away."

Walton then left the "exam room" via a long, narrow hallway, which led to a spherical room with only a high-backed chair placed in the center of the room. Though he was afraid there might be someone seated in the chair, Walton says he walked towards it. As he did, lights began to appear in the room. The chair was empty, so Walton says he sat in it. When he did, the room was suddenly filled with lights, similar to stars projected on a round planetarium ceiling and the floor under him seemed to vanish, making it appear as if he was floating in space.

The chair was equipped on the left arm with a single short thick lever with an oddly shaped molded handle atop some dark brown material. On the right arm, there was an illuminated, lime-green screen about five inches square with black lines intersected at all angles.

When Walton pushed the lever, he reported that the stars rotated around him slowly. When he released the lever, the stars remained at their new position. He decided to stop manipulating the lever, since he had no idea what it might do.

He left the chair, and the stars disappeared. Walton thought he had seen a rectangular outline on the rounded wall— perhaps a door— and went to look for it.

Just then, Walton heard a sound behind him. He turned, expecting more of the short, large eyed creatures, but was pleasantly surprised to see a tall being, representing a tall human about six feet, two inches in height. He was extremely muscular and evenly proportioned. He appeared to weigh about two hundred pounds. He was wearing blue coveralls, his feet were covered with black boots, a black band or belt wrapped around his middle. He carried no tools or weapons on his belt or in his hands; no insignia marked his clothing. However, he did wear a glassy round helmet on his head. At the time, Walton said, he did not realize how odd the man's eyes were: larger than normal, and a bright golden in color.

Walton says he then asked the man a number of questions, but the man only grinned and motioned for Walton to follow him. Walton also said that because of the man's helmet he might have been unable to hear him, so he followed the man down a hallway which led to a door and a steep ramp down to a large room Walton described as similar to an aircraft hangar. Walton says he realized he had just left a disc-shaped craft similar to the one he had seen in the forest just before he had been struck by the beam of light, but the craft was perhaps twice as large.

In the hangar-like room, Walton reported seeing other disc-shaped craft. The man led him to another room, containing three more human like beings— a woman and two men— resembling the helmeted man. These people did not wear glass helmets, so Walton says he began asking them questions. They responded with the same dull grin, and gently led him by his arm to a small table.

Once he was seated on the table, Walton says he realized the woman held a device like an oxygen mask, only there were no tubes connected to it. The only thing attached to it was a small black golfball-sized sphere. She placed the mask on his face. Before he could fight back, Walton says everything went black and he immediately passed out.

168

When he woke again, Walton says he was outside the gas station in Heber, Arizona. One of the disc-shaped craft was hovering just above the highway. After a moment, the craft shot away, and Walton stumbled to the telephones and called his brother-in-law, Grant Neff. He thought that only a few hours had passed.

After hearing Walton's story, Gillespie speculated that Walton may have been hit on the head and drugged, then taken to a normal hospital where he had confused the details of a routine exam with something more spectacular. Walton dismissed this, noting that the medical examination had found no trace of head trauma or drugs in his system. Walton told Sheriff Gillespie that he was willing to take a polygraph, a truth serum, or undergo hypnosis to support his account. Gillespie said that a polygraph would suffice, and he promised to arrange one in secret to avoid the growing media circus.

Duane and Travis then drove to Scottsdale, Arizona, where a meeting with APRO consultant James A. Harder had been arranged. Harder hypnotized Walton, hoping to uncover more details of the missing five days. Clark writes that "Unlike many other abductees, however, Walton's conscious recall and unconscious 'memory' were the same, and he could account for only a maximum of two hours, and perhaps less, of his missing five days. Curiously … Walton encountered an impenetrable mental block and expressed the view that he would 'die' if the regression continued."

01-06-1976 Stanford Abduction. Three women claimed to be abducted after seeing a bright red object and a blue light while driving in a car on the highway.

06-77-1976 1976 Canary Isles sightings. Several lights and a spherical transparent blue craft, piloted by two beings was reported.

08-26-1976 Allagash Abductions. Brothers Jim Weiner and Jack Weiner with friends Charles Foltz and Charles Rak claim that they were abducted by aliens during a camping trip in Allagash, Maine on August 20, 1976. According to the four men, hypnotic regression enabled them to recall being taken aboard a circular UFO and being "probed and tested by four-fingered beings with almond-shaped eyes and languid limbs".

The incident was the subject of a book, *The Allagash Abductions* by Raymond E. Fowler. The incident was also dramatized in an episode of *Unsolved Mysteries* and featured in the TV program *Abducted by UFOs*.

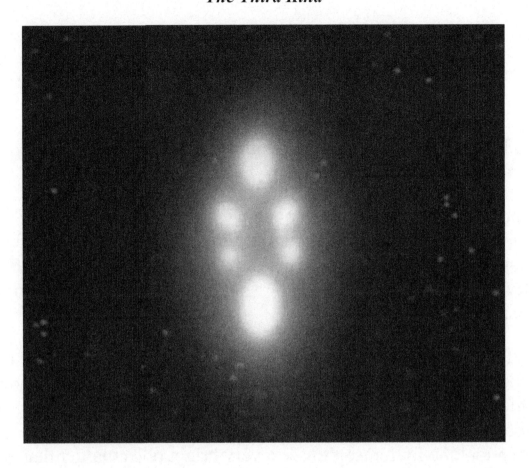

09-19-1976 The **1976 Tehran UFO Incident** was a radar and visual sighting of an unidentified flying object (UFO) over Tehran, the capital of Iran, during the early morning hours of 19 September 1976. During the incident, two F-4 Phantom II jet interceptors reported losing instrumentation and communications as they approached, only to have them restored upon withdrawal; one of the aircraft also reported suffering temporary weapons systems failure, while preparing to open fire.

The incident, recorded in a four-page U.S. Defense Intelligence Agency (DIA) report distributed to at least the White House, Secretary of State, Joint Chiefs of Staff, National Security Agency (NSA) and Central Intelligence Agency (CIA), remains one of the most well-documented military encounters with anomalous phenomena in history, and various senior Iranian military officers directly involved with the events have gone on public record stating their belief that the object was not of terrestrial origin.

At approximately 0030 hours local time (2100Z), 19 September 1976, the Imperial Iranian Air Force command post at Tehran received four reports by telephone, from civilians in the Shemiran city district, of unusual activity in the night sky. The callers reported seeing an object similar to a star, but much brighter.

When the command post found no helicopters airborne to account for the reports, they called General Yousefi, assistant deputy commander of operations. General Yousefi at first said the object was only a star, but after conferring with the control tower at Mehrabad International Airport and then looking for himself to see a very bright object larger than a star, he decided to scramble one F-4 Phantom II jet fighter from Shahrokhi Air Force Base in Hamadan, approximately 175 miles (282 km) west of Tehran (for location see map at right).

At 0130 hours (2200Z), the F-4, piloted by Captain Mohammad Reza Azizkhani was launched and proceeded to a point 40 nautical miles (74 km) north of Tehran. It was noted that the object was of such brilliance that it could be seen from 70 miles (110 km) away. When the aircraft approached to approximately 25 nautical miles (46 km) from the object, the jet lost all instrumentation and communications capabilities, prompting Azizkhani to break off the intended intercept and turn back toward Shahrokhi; upon the evasion, both systems resumed functioning.

At 0140 hours, a second F-4 was scrambled, piloted by Lieutenant Parviz Jafari and Lieutenant Jalal Damirian. Jafari would eventually retire as a general and participate on 12 November 2007, at a National Press Club conference demanding a worldwide investigation into UFO phenomena. (see also below) Jafari's jet had acquired a radar lock on the object at 27 nautical miles (50 km) range. The radar signature of the UFO resembled that of a Boeing 707 aircraft. Closing on the object at 150 nautical miles (280 km) per hour and at a range of 25 nautical miles (46 km), the object began to move, keeping a steady distance of 25 nautical miles (46 km) from the F-4. The size of the object was difficult to determine due to its intense brilliance. The lights of the object were alternating blue, green, red, and orange, and were arranged in a square pattern. The lights flashed in sequence, but the flashing was so rapid that they all could be seen at once.

While the object and the F-4 continued on a southerly path, a smaller second object detached itself from the first and advanced on the F-4 at high speed. Lieutenant Jafari, thinking he was under attack, tried to launch an AIM-9 sidewinder missile, but he suddenly lost all instrumentation, including weapons control, and all communication. He later stated he attempted to eject, but to no avail, as this system, which is entirely mechanical, also malfunctioned. Jafari then instituted a turn and a negative G dive as evasive action. The object fell in behind him at about 3 to 4 nautical miles (7.4 km) distance for a short time, then turned and rejoined the primary object.

Once again, as soon as the F-4 had turned away, instrumentation and communications were regained. The F-4 crew then saw another brightly lit object detach itself from the other side of the primary object and drop straight down at high speed. The F-4 crew expected it to impact the ground and explode, but it came to rest gently.

The F-4 crew then overflew the site at a decreased altitude and marked the position of the light's touchdown. Jafari would later comment that the object was so bright that it lit up the ground and he could see rocks around it. The object had touched down near Rey Oil Refinery on the outskirts of Iran. Then they landed at Mehrabad, noting that each time they passed through a magnetic bearing of 150 degrees from Mehrabad, they experienced interference and communications failure.

A civilian airliner that was approaching Mehrabad also experienced a loss of communications at the same position relative to Mehrabad. As the F-4 was on final approach, they sighted yet another object, cylinder-shaped, with bright, steady lights on each end and a flashing light in the middle. The object overflew the F-4 as they were on approach. Mehrabad tower reported no other aircraft in the area, but tower personnel were able to see the object when given directions by Jafari. Years later, the main controller and an investigating general revealed that the object also overflew the control tower and knocked out all of its electronic equipment as well (see below).

The next day, the F-4 crew flew out in a helicopter to the site where they had seen the smaller object land. In the daylight, it was determined to be a dry lake bed, but no traces could be seen. They then circled the area to the west and picked up a noticeable "beeper" signal. The signal was loudest near a small house, so they landed and questioned the occupants of the house about any unusual events of the previous night. They reported a loud noise and a bright light like lightning.

Further investigation of the landing site, including radiation testing of the area was apparently done, but the results were never made public. Since this event occurred before the fall of the Shah, any records in Tehran may be lost.

1977 Colares UFO flap. Claims of UFOs on a river island investigated by the air force in "Operação Prato" and civilian researchers, among them J. F. Vallée.

1977 Broad Haven. In the 1970s, the area was the scene of alleged UFO sightings and nicknamed the Broad Haven Triangle.

05-07-1978 Emilcin Abduction. Jan Wolski (/ˈvɒlski/; 29 May 1907 – 8 January 1990) was out driving a horse-drawn cart early on 10 May 1978 when he says he was jumped by two "short, green-faced humanoid entities" about 5 feet (1.5 m) tall. The two beings jumped onto Wolski's cart and, according to Wolski, sat next to him and started to speak in a strange language. Originally he had mistaken them for foreigners because of their "slanted eyes and prominent cheekbones." Wolski drove his cart, with the two beings aboard, to a clearing where he says a large object was hovering.

According to Wolski, a purely white unidentified flying object, about 14.75 feet (4.5 m) – 16.5 feet (5.0 m) in height and "as long as a bus," hovered in the air at an altitude of about 16 feet (4.9 m). There were no notable external features of the craft (i.e. lights, joints, etc.). Wolski mentioned that there were four objects on the craft made of a black material that appeared to be drill-like in appearance, which generated a humming sound. An elevator-like platform attached to the hovering craft descended to the ground. (Artist's Rendition 1 2)

Wolski then claims that he was taken aboard the ship with two additional entities he met near the flying object. He was then gestured to "dress down" (take off his clothes). There were about eight or ten benches situated around the craft, each the size for one person to sit in. There were some rooks in front of the door which was moving its legs and wings but seemed to be immobilizied. Wolski claims that he was then examined with a tool that resembled two dishes or "saucers." After this, he was ordered to redress, and it was then that he noticed there were no lights or windows on the craft, only the daylight coming through the craft's door. Entities ate and offered him something like icicles but he refused. The craft's interior was described as black with a greyish tint, similar to that of the creatures' outfits.

Afterwards, Wolski returned home to his family and notified them of what had just happened, urging them to come see the floating craft. He notified his sons who called to other neighbors, and together they went to investigate the site. The grass where the craft had been had signs of usage in it, being "trodden down covered with dew and paths coming in all directions." Wolski returned home, leaving the rest of the neighbors and family at the site.

Wolski's sons claim that there were footprints left behind by the beings, though they did not detail whether the footprints were larger or smaller than their own.

A six-year-old boy claims to have witnessed a bus-like craft hover over a barn, then climb high into the sky and vanish.

Wolski explained these memories in an interview with Henryk Pomorski and Krystyna Adamczyk in July 1978, two months following the incident. The audio tape of the interview was kept in a private archive for a long time before being released to the public.

In 2005, a memorial was constructed in Emilcin to commemorate the alien abduction of Jan Wolski. The text in Polish, read: "On 10 May 1978 in Emilcin a UFO object landed. The truth will astonish us in the future".

10-21-1978 Valentich disappearance. Twenty-year-old **Frederick Valentich disappeared** while on a 125-mile (235 km) training flight in a Cessna 182L light aircraft over Bass Strait in Australia on 21 October 1978.

Described as a "flying saucer enthusiast", Valentich radioed Melbourne air traffic control that he was being accompanied by an aircraft about 1,000 feet (300 m) above him, that his engine had begun running roughly, and finally reported, "It's not an aircraft."

There were belated reports of a UFO sighting in Australia on the night of the disappearance; however, Associated Press reported that the Department of Transport was skeptical a UFO was behind Valentich's disappearance, and that some of their officials speculated that

"Valentich became disorientated and saw his own lights reflected in the water, or lights from a nearby island, while flying upside down."

Frederick Valentich had about 150 total hours' flying time and held a class-four instrument rating, which authorised him to fly at night, but only "in visual meteorological conditions". He had twice applied to enlist in the Royal Australian Air Force but was rejected because of inadequate educational qualifications. He was a member of the Air Training Corps, determined to have a career in aviation. Valentich was studying part-time to become a commercial pilot but had a poor achievement record, having twice failed all five commercial license examination subjects, and as recently as the previous month had failed three more commercial license subjects. He had been involved in flying incidents, for example, straying into a controlled zone in Sydney, for which he received a warning, and twice deliberately flying into a cloud, for which prosecution was being considered. According to his father, Guido, Frederick was an ardent believer in UFOs and worried about attacks from UFOs.

Valentich radioed Melbourne Flight Service at 7:06 PM to report an unidentified aircraft was following him at 4,500 feet (1,400 m) and was told there was no known traffic at that level. Valentich said he could see a large unknown aircraft which appeared to be illuminated by four bright landing lights. He was unable to confirm its type, but said it had passed about 1,000 feet (300 m) overhead and was moving at high speed. Valentich then reported that the aircraft was approaching him from the east and said the other pilot might be purposely toying with him. Valentich said the aircraft was "orbiting" above him and that it had a shiny metal surface and a green light on it.

Valentich reported that he was experiencing engine problems. Asked to identify the aircraft, Valentich radioed, "It isn't an aircraft" when his transmission was interrupted by unidentified noise described as being "metallic, scraping sounds" before all contact was lost.

A sea and air search was undertaken that included oceangoing ship traffic, a P-3 Orion aircraft, plus eight civilian aircraft. The search encompassed over 1,000 square miles. Search efforts ceased on 25 October 1978.

A Department of Transport (DOT) investigation into Valentich's disappearance was unable to determine the cause, but that it was "presumed fatal" for Valentich. Five years after Valentich's plane went missing, an engine cowl flap was found washed ashore on Flinders Island. In July 1983, the Bureau of Air Safety Investigation asked The Royal Australian Navy Research Laboratory (RANRL) about the likelihood that the cowl flap might have "traveled" to its ultimate position from the region where the plane disappeared. The bureau noted that "the part has been identified as having come from a Cessna 182 aircraft between a certain range of serial numbers" which included Valentich's aircraft. The bureau also noted that while it is possible for cowl flaps to separate from aircraft in flight, this had not happened with any recent aircraft.

It has been proposed that Valentich staged his own disappearance: even taking into account a trip of between 30 and 45 minutes to Cape Otway, the aircraft still had enough fuel to fly 800 kilometers; despite ideal conditions, at no time was the aircraft plotted on radar, casting doubts as to whether it was ever near Cape Otway;

and Melbourne Police received reports of a light aircraft making a mysterious landing not far from Cape Otway at the same time as Valentich's disappearance.

Another proposed explanation is that Valentich became disoriented and was flying upside down. What he thought he saw, if this were the case, would be his own aircraft's lights reflected in the water. He would then have crashed into the water.

Another proposed possibility is suicide, although it has been suggested that he had a contented lifestyle.

A 2013 review of the radio transcripts and other data by astronomer and retired U.S. Air Force pilot James McGaha and author Joe Nickell proposes that the inexperienced Valentich was deceived by the illusion of a tilted horizon for which he attempted to compensate and inadvertently put his plane into a downward, so-called "graveyard spiral" which he initially mistook for simple orbiting of the plane. According to the authors, the G-forces of a tightening spiral would decrease fuel flow, resulting in the "rough idling" reported by the pilot. McGaha and Nickell also propose that the apparently stationary, overhead lights that Valentich reported were probably the planets Venus, Mars and Mercury, along with the bright star Antares, which would have behaved in a way consistent with the pilot's description.

UFOlogists have speculated that extraterrestrials either destroyed Valentich's plane or abducted him, asserting that some individuals reported seeing "an erratically moving green light in the sky" and that he was "in a steep dive at the time." Ufologists believe these accounts are significant because of the "green light" mentioned in Valentich's radio transmissions.

Phoenix, Arizona-based UFO group Ground Saucer Watch claim that photos taken that day by plumber Roy Manifold show a fast-moving object exiting the water near Cape Otway lighthouse. Though the pictures were not clear enough to identify the object, UFO groups argue that they show "a bona fide unknown flying object, of moderate dimensions, apparently surrounded by a cloud-like vapor/exhaust residue."

12-21-1978 Kaikoura lights. The **Kaikoura lights** is a name given by the New Zealand media to a series of sightings that occurred in December 1978, over the skies above the Kaikoura mountain ranges in the northeast of New Zealand's South Island. The first sightings were made on 21 December when the crew of a Safe Air Ltd cargo aircraft began observing a series of strange lights around their Armstrong Whitworth AW.660 Argosy aircraft, which tracked along with their aircraft for several minutes before disappearing and then reappearing elsewhere, the UFO was very large and had five white flashing lights that were visible on the craft. Some people say that they could see some little disks drop from the UFO and then disappear (they were never found). The pilots described some of the lights to be the size of a house and others small but flashing brilliantly. These objects appeared on the air traffic controller radar in Wellington and also on the aircraft's on-board radar.

On 30 December 1978, a television crew from Australia recorded background film for a network show of interviews about the sightings. For many minutes at a time on the flight to Christchurch, unidentified lights were observed by five people on the flight deck, were tracked by Wellington Air Traffic Controllers, and filmed in color by the television crew. One object reportedly followed the aircraft almost until landing.

The cargo plane then took off again with the television crew still on board, heading for Blenheim. When the aircraft reached about 2000 feet, it encountered a gigantic lighted orb which fell into station off the wing tip and tracked along with the cargo aircraft for almost quarter of an hour, while being filmed, watched, tracked on the aircraft radar and described on a tape recording made by the TV film crew.

A spate of sightings followed the initial report and an Air Force Skyhawk was put on stand-by to investigate any positive sightings.

01-03-1979 Mindalore Incident. A mother and her 12-year-old son of Mindalore, Krugersdorp, both claimed an encounter with a group of human-like entities standing beside a craft. One of them encouraged her to depart with them permanently. After she refused, they entered the craft which then shot upwards and disappeared in 30 seconds.

08-27-1979 Val Johnson incident. Johnson reported that while he was on patrol near Stephen, Minnesota about 2 AM on August 27, 1979 he saw a beam of light just above the road. According to Johnson, the beam sped towards him, his squad car was engulfed in light, and he heard glass breaking. Johnson said he was unconscious for 39 minutes and when he awoke he realized his wristwatch and the vehicle's clock had stopped for 14 minutes. The windshield was shattered, a headlight and red emergency light was damaged and a thin radio aerial bent. Deputies responding to Johnson's call for help found the squad car sideways on the road. Johnson suffered bruises and eye irritation that a physician compared to "welder's burns". When the story received national publicity, Johnson told reporters the sudden attention had caused him and his family a great deal of emotional strain. On September 11, 1979, Johnson appeared as a guest on ABC's *Good Morning America* program.

UFOlogists consider the incident one of the most significant and best-publicized UFO events of the 1970s. Allan Hendry of the Center for UFO Studies investigated the damage to Johnson's car along with other aspects of the incident and concluded that Johnson had not hoaxed the event. According to UFOlogist Jerome Clark, Johnson refused to take a polygraph test because he felt that doing so "would only satisfy people's morbid curiosity". In his 1983 book *UFOs: The Public Deceived*, UFO skeptic Philip Klass argued that the entire event was a hoax, and that Johnson had deliberately damaged his own patrol car.

11-09-1979 Dechmont Woods Encounter. In ufology, the **Robert Taylor Incident**, aka **Livingston Incident** or **Dechmont Woods Encounter** is the name given to claims of sighting an extraterrestrial spacecraft on Dechmont Law in Livingston, West Lothian, Scotland in 1979 by forester Robert "Bob" Taylor. When Taylor returned home from a trip to Dechmont Law dishevelled, his clothes torn and with grazes to his chin and thighs, he claimed he'd encountered a "flying dome" which tried to pull him aboard. Due to his injuries, the police recorded the matter as a common assault and the incident is popularly promoted as the "only example of an alien sighting becoming the subject of a criminal investigation".

According to Taylor, a forestry worker for the Livingston Development Corporation, on 9 November 1979, he parked his pickup truck at the side of a road near the M8 motorway and walked along a forest path up the side of Dechmont Law with his dog.

Taylor reported seeing what he described as a "flying dome" or a large, circular sphere approximately 7 yards (6.4 meters) in diameter, hovering above the forest floor in a clearing about 530 yards (480 meters) away from his truck. Taylor described the object as "a dark metallic material with a rough texture like sandpaper" featuring an outer rim "set with small propellers".

Taylor claims he experienced a foul odor "like burning brakes" and that smaller spheres "similar to sea mines" had seized him and were dragging him in the direction of the larger object when he lost consciousness. According to Taylor, he later awoke and the objects were gone, but he could not start his truck, so he walked back to his home in Livingston.

Taylor's wife reported that when he arrived home on foot, he appeared dishevelled and muddy with torn clothing and ripped trousers. His wife called the police and a doctor, who treated him for grazes to his chin and thighs. Police accompanied Taylor to the site where he claimed he received his injuries. They found "ladder-shaped marks" in the ground where Taylor said he saw the large spherical object and other marks that Taylor said were made by the smaller, mine-like objects. Police recorded the matter as a criminal assault.

The story drew attention from ufologists, who erected a plaque on the site of the alleged encounter, and Taylor became notable among UFO enthusiasts for being involved in the only UFO sighting that was subject to a criminal investigation. Ufologist and author Malcolm Robinson accepts Taylor's story, saying he believes "it could be one of the few genuine cases of a UFO encounter".

In 1979, the UFO sceptic Steuart Campbell visited the scene of the incident with the police. Campbell was convinced that a simple explanation would be found. On his second visit to the site he stated that he had observed some PVC pipes in an adjoining field. He discovered that the local water authority had laid a cable duct within 100m of the clearing. He came to the conclusion that stacks of pipes may have been stored in the clearing and were responsible for the ground markings.

Patricia Hannaford, founder of the Edinburgh University UFO Research Society and a qualified physician advised Campbell on medical aspects of the case. She suggested that Taylor's collapse was an isolated attack of temporal lobe epilepsy, and the fit explained the objects as hallucinations. Symptoms such as Taylor's previous meningitis, his report of a strong smell which nobody else could detect, his headache, dry throat, paralysis of his legs and period of unconsciousness suggested this cause.

Steve Donnelly a physicist and editor for *The Skeptic* also considered the incident to be explained by an epileptic attack. Campbell suggested Taylor's attack may have been stimulated by a mirage of Venus.

Local businessman Phill Fenton published a report in 2013 speculating that Taylor "may have suffered a mini-stroke and been exposed to harmful chemicals which left him confused and disoriented" and that "the UFO he believes he saw could have been a saucer-shaped water tower nearby".

11-11-1979 The **Manises UFO incident** took place on 11 November 1979, forcing a commercial flight to make an emergency landing at the Manises' airport in Valencia, Spain.

A TAE's (former airline) Supercaravelle was the first aircraft involved in the incident. Flight JK-297 had taken off from Salzburg (Austria) with 109 passengers on board, and had made a refuelling stop on the island Mallorca before setting course towards Las Palmas.

Halfway through the flight, at about 23:00h, Pilot Francisco Javier Lerdo de Tejada and his crew noticed a set of red lights that were fast approaching the aircraft. These lights appeared to be on a collision course with the aircraft, alarming the crew. The captain requested information about the inexplicable lights, but neither the military radar of Torrejón de Ardoz (Madrid) nor the flight control center in Barcelona could provide any explanation for this phenomenon.

In order to avoid a possible collision, the captain changed altitude. However, the lights mirrored the new course and stayed about half a kilometer away from the plane. Since the object was violating all elementary safety rules and an evasive maneuver was deemed impossible by the crew, the captain decided on going off-course and made an emergency landing in Manises' airport. This was the first time in history in which a commercial flight was forced to make an emergency landing because of a UFO.

The flight crew reported the lights abandoning the pursuit just before the landing took place. However, three new UFO signals were detected by the radar, each one with an estimated diameter of 200 meters. The objects were seen by several witnesses.

One of the UFOs passed very close to the airport runway, and emergency lights were lit by the land crew in case the object happened to be an unregistered flight experiencing difficulties.

Given the lack of answer to all communication attempts, a Mirage F-1 took off from the nearby airbase of Los Llanos (Albacete) to identify the phenomenon. The pilot, Spanish Air Force captain Fernando Cámara, had to increase his speed to mach 1.4 just to be able to get visual contact with what he perceived to be a truncated cone shape displaying a changing bright color, but despite his initial efforts the object quickly disappeared from sight. The pilot was informed of a new radar echo, which indicated that another object might be near Sagunto (Valencia).

When the pilot was close enough, the object accelerated and disappeared again. This time, though, the UFO seemed to respond and the fighter had its avionics scrambled -its electronic flight systems were jammed. At last, and after a third contact attempt, the UFO finally disappeared, heading for Africa. After an hour and a half of pursuit, and due to fuel shortage, the pilot was forced to return to the base with no results.

12-24-1980 to 12-28-1980 Rendlesham Forest Incedent. In late December 1980, there were a series of reported sightings of unexplained lights near Rendlesham Forest, Suffolk, England, which have become linked with claims of UFO landings. The events occurred just outside RAF Woodbridge, which was used at the time by the U.S. Air Force. USAF personnel including deputy base commander Lieutenant Colonel Charles I. Halt claimed to see things they described as a UFO sighting.

The occurrence is the most famous of claimed UFO events to have happened in Britain, ranking among the best-known purported UFO events worldwide. It has been compared to the Roswell UFO incident in the United States and is sometimes referred to as "Britain's Roswell". The Ministry of Defence (MoD) stated the event posed no threat to national security, and it therefore never was investigated as a security matter. The sightings have been explained as a misinterpretation of a series of nocturnal lights – a fireball, the Orford Ness lighthouse and bright stars.

26 December Around 3:00 a.m. on 26 December 1980 (reported as the 27th by Halt in his memo to the UK Ministry of Defence – see below) a security patrol near the east gate of RAF Woodbridge saw lights apparently descending into nearby Rendlesham Forest. These lights have been attributed by astronomers to a piece of natural debris seen burning up as a fireball over southern England at that time. Servicemen initially thought it was a downed aircraft but, upon entering the forest to investigate they saw, according to Halt's memo, what they described as a glowing object, metallic in appearance, with colored lights. As they attempted to approach the object, it appeared to move through the trees, and "the animals on a nearby farm went into a frenzy". One of the servicemen, Sergeant Jim Penniston, later claimed to have encountered a "craft of unknown origin" while in the forest although there was no mention of this at the time and there is no corroboration from other witnesses.

Shortly after 4:00 a.m. local police were called to the scene but reported that the only lights they could see were those from the Orford Ness lighthouse, some miles away on the coast.

After daybreak on the morning of 26 December, servicemen returned to a small clearing near the eastern edge of the forest and found three small impressions in a triangular pattern, as well as burn marks and broken branches on nearby trees.

At 10.30 a.m. the local police were called out again, this time to see the impressions on the ground, which they thought could have been made by an animal. Georgina Bruni, in her book *You Can't Tell the People*, published a photograph of the supposed landing site taken on the morning after the first sighting.

28 December The deputy base commander Lt Col Charles Halt visited the site with several servicemen in the early hours of 28 December 1980 (reported as the 29th by Halt). They took radiation readings in the triangle of depressions and in the surrounding area using an AN/PDR-27, a standard US military radiation survey meter. The significance of the readings they obtained is disputed. Halt recorded the events on a micro-cassette recorder (see "The Halt Tape", below).

It was during this investigation that a flashing light was seen across the field to the east, almost in line with a farmhouse, as the witnesses had seen on the first night. The Orford Ness lighthouse is visible further to the east in the same line of sight (see below).

Later, according to Halt's memo, three starlike lights were seen in the sky, two to the north and one to the south, about 10 degrees above the horizon. Halt said that the brightest of these hovered for two to three hours and seemed to beam down a stream of light from time to time. Astronomers have explained these starlike lights as bright stars.

DEPARTMENT OF THE AIR FORCE
LFO ROW NOT CAG

CO   13 Jan 81

Unexplained Lights

RAF/CC

1. Early in the morning of 27 Dec 80 (approximately 0300L), two USAF
security police patrolmen saw unusual lights outside the back gate at
RAF Woodbridge. Thinking an aircraft might have crashed or been forced
down, they called for permission to go outside the gate to investigate.
The on-duty flight chief responded and allowed three patrolmen to pro-
ceed on foot. The individuals reported seeing a strange glowing object
in the forest. The object was described as being metalic in appearance
and triangular in shape, approximately two to three meters across the
base and approximately two meters high. It illuminated the entire forest
with a white light. The object itself had a pulsing red light on top and
a bank(s) of blue lights underneath. The object was hovering or on legs.
As the patrolmen approached the object, it maneuvered through the trees
and disappeared. At this time the animals on a nearby farm went into a
frenzy. The object was briefly sighted approximately an hour later near
the back gate.

2. The next day, three depressions 1 1/2" deep and 7" in diameter were
found where the object had been sighted on the ground. The following
night (29 Dec 80) the area was checked for radiation. Beta/gamma readings
of 0.1 milliroentgen were recorded with peak readings in the three depressions and near the center of the triangle formed by the depressions.
A nearby tree had moderate (.05-.07) readings on the side of the tree
toward the depressions.

3. Later in the night a red sun-like light was seen through the trees.
It moved about and pulsed. At one point it appeared to throw off glowing
particles and then broke into five separate white objects and then disappeared. Immediately thereafter, three star-like objects were noticed
in the sky, two objects to the north and one to the south, all of which
were about 10° off the horizon. The objects moved rapidly in sharp angular
movements and displayed red, green and blue lights. The objects to the
north appeared to be elliptical through an 8-12 power lens. They then
turned to full circles. The objects to the north remained in the sky for
an hour or more. The object to the south was visible for two or three
hours and beamed down a stream of light from time to time. Numerous individuals, including the undersigned, witnessed the activities in paragraphs
2 and 3.

CHARLES I. HALT, Lt Col, USAF
Deputy Base Commander

192

The Halt memo

The first piece of primary evidence to be made available to the public was a memorandum written by the deputy base commander, Lt. Col. Charles I. Halt, to the Ministry of Defence (MoD). Known as the "Halt memo", this was made available publicly in the United States under the US Freedom of Information Act in 1983. The memorandum (left), was dated "13 Jan 1981" and headed "Unexplained Lights". The two-week delay between the incident and the report might account for errors in dates and times given. The memo was not classified in any way. Dr David Clarke, a consultant to the National Archives, has investigated the background to this memo and the reaction to it at the Ministry of Defence. His interviews with the personnel involved confirmed the cursory nature of the investigation made by the MoD, and failed to find any evidence for any other reports on the incident made by the USAF or UK apart from the Halt memo. Halt has since gone on record as saying he believes that he witnessed an extraterrestrial event that was then covered up.

The Halt Tape

In 1984, a copy of what became known as the "Halt Tape" was released to UFO researchers by Col Sam Morgan, who had by then succeeded Ted Conrad as Halt's superior. This tape chronicles Halt's investigations in the forest in real time, including taking radiation readings, the sighting of the flashing light between trees and the starlike objects that hovered and twinkled. The tape has been transcribed by researcher Ian Ridpath, who includes a link to an audio download and also a step-by-step analysis of the entire contents of the tape.

Michael Ryan

The Halt affidavit

In June 2010, retired colonel Charles Halt signed a notarised affidavit, in which he again summarised what had happened, then stated he believed the event to be extraterrestrial and it had been covered up by both the UK and US. Contradictions between this affidavit and the facts as recorded at the time in Halt's memo and tape recording have been pointed out.

In 2010, base commander Colonel Ted Conrad provided a statement about the incident to Clarke. Conrad stated that "We saw nothing that resembled Lieutenant Colonel Halt's descriptions either in the sky or on the ground" and that "We had people in position to validate Halt's narrative, but none of them could." In an interview, Conrad criticised Halt for the claims in his affidavit, saying "he should be ashamed and embarrassed by his allegation that his country and Britain both conspired to deceive their citizens over this issue. He knows better." Conrad also disputed the testimony of Sergeant Jim Penniston, who claimed to have touched an alien spacecraft; he said that he interviewed Penniston at the time and he had not mentioned any such occurrence. Conrad also suggested that the entire incident might have been a hoax.

A 1983 Omni article says "Colonel Ted Conrad the base commander... recalls five Air Force policemen spotted lights from what they thought was a small plane descending into the forest. Two of the men tracked the object on foot and came upon a large tripod-mounted craft. It had no windows but was studded with brilliant red and blue lights. Each time the men came within 50 yards of the ship, Conrad relates, it levitated six feet in the air and backed away. They followed it for almost an hour through the woods and across a field until it took off at 'phenomenal speed.' Acting on the reports made by his men, Colonel Conrad began a brief investigation of the incident in the morning.

He went into the forest and located a triangular pattern ostensibly made by the tripod legs. ...he did interview two of the eyewitnesses and concludes, 'Those lads saw something, but I don't know what it was'."

Ministry of Defense file

Evidence of a substantial MoD file on the subject led to claims of a cover-up; some interpreted this as part of a larger pattern of information suppression concerning the true nature of unidentified flying objects, by both the United States and British governments. However, when the file was released in 2001 it turned out to consist mostly of internal correspondence and responses to inquiries from the public. The lack of any in-depth investigation in the publicly released documents is consistent with the MoD's earlier statement that they never took the case seriously. Included in the released files is an explanation given by defence minister Lord Trefgarne as to why the MoD did not investigate further.

In 2005, the Forestry Commission used Lottery proceeds to create a trail in Rendlesham Forest because of public interest and nicknamed it the *UFO Trail*. In 2014, the Forestry Service commissioned an artist to create a work which has been installed at the end of the trail. The artist states the piece is modeled after sketches that purportedly represent some versions of the UFO claimed to have been seen at Rendlesham.

12-29-1980 The **Cash-Landrum Incident** was a reported Unidentified Flying Object sighting from the United States in 1980, which witnesses insist was responsible for damage to their health. It is one of very few UFO cases to result in civil court proceedings.

It can be classified as a Close Encounter of the Second Kind, due to its reported physical effects on the witnesses and their automobile.

Skeptical ufologist Peter Brookesmith wrote:

> "To ufologists, the case is perhaps the most baffling and frustrating of modern times, for what started with solid evidence for a notoriously elusive phenomenon petered out in a maze of dead ends, denials, and perhaps even official deviousness."

The Third Kind

On the evening of December 29, 1980, Betty Cash, Vickie Landrum and Colby Landrum (Vickie's seven-year-old grandson) were driving home to Dayton, Texas in Cash's Oldsmobile Cutlass after dining out.

At about 9:00 p.m., while driving on an isolated two-lane road in dense woods, the witnesses said they observed a light above some trees. They initially thought the light was an airplane approaching Houston Intercontinental Airport (about 35 miles away) and gave it little notice.

A few minutes later on the winding roads, the witnesses saw what they believed to be the same light as before, but it was now much closer and very bright. The light, they claimed, came from a huge diamond-shaped object, which hovered at about treetop level. The object's base was expelling flame and emitting significant heat.

Vickie Landrum told Cash to stop the car, fearing they would be burned if they approached any closer. However, Vickie's opinion of the object quickly changed: a born again Christian, she interpreted the object as a sign of the second coming of Jesus Christ, telling her grandson, "That's Jesus. He will not hurt us." (Clark, 175)

Anxious, Cash considered turning the car around, but abandoned this idea because the road was too narrow and she presumed the car would get stuck on the dirt shoulders, which were soft from that evening's rains.

Cash and Landrum got out of the car to examine the object. Colby was terrified, however, and Vickie Landrum quickly returned to the car to comfort the frantic child. Cash remained outside the car, "mesmerized by the bizarre sight," as Jerome Clark wrote. (Clark, 175) He went on,

The object, intensely bright and a dull metallic silver, was shaped like a huge upright diamond, about the size of the Dayton water tower, with its top and bottom cut off so that they were flat rather than pointed. Small blue lights ringed the center, and periodically over the next few minutes flames shot out of the bottom, flaring outward, creating the effect of a large cone. Every time the fire dissipated, the UFO floated a few feet downwards toward the road. But when the flames blasted out again, the object rose about the same distance." (Clark, 175)

The witnesses said the heat was strong enough to make the car's metal body painful to the touch—Cash said she had to use her coat to protect her hand from being burnt when she finally re-entered the car. When she touched the car's dashboard, Vickie Landrum's hand pressed into the softened vinyl, leaving an imprint that was evident weeks later. Investigators cited this handprint as proof of the witnesses' account; however, no photograph of the alleged handprint exists.

The object then moved to a point higher in the sky. As it ascended over the treetops, the witnesses claimed that a group of helicopters approached the object and surrounded it in tight formation. Cash and Landrum counted 23 helicopters, and later identified some of them as tandem-rotor CH-47 Chinooks used by military forces worldwide.

With the road now clear, Cash drove on, claiming to see glimpses of the object and the helicopters receding into the distance.

From first sighting the object to its departure, the witnesses said the encounter lasted about 20 minutes. Based on descriptions given in John F. Schuessler's book about the incident, it appears that the observers were southbound on Texas state highway FM 1485/2100 when they claimed to have seen the object. The initial location of the reported object, based on the same descriptions, was just south of Inland Road.

After the UFO and helicopters left, Cash took the Landrums home, then retired for the evening. That night, they all experienced similar symptoms, though Cash to a greater degree. All suffered from nausea, vomiting, diarrhea, generalized weakness, a burning sensation in their eyes, and feeling as though they'd suffered sunburns.

Over the next few days, Cash's symptoms worsened, with many large, painful blisters forming on her skin. When taken to a hospital emergency room on January 3, 1981, Clark writes, Cash "could not walk, and had lost large patches of skin and clumps of hair. She was released after 12 days, though her condition was not much better, and she later returned to the hospital for another 15 days."(Clark, 176)

The Landrums' health was somewhat better, though both suffered from lingering weakness, skin sores and hair loss.

A radiologist who examined the witnesses' medical records for MUFON wrote, "We have strong evidence that these patients have suffered secondary damage to ionizing radiation. It is also possible that there was an infrared or ultraviolet component as well." (quoted in Clark, 176)

However, Brad Sparks contends that, although the symptoms were somewhat similar to those caused by ionizing radiation, the rapidity of onset was only consistent with a massive dose that would have meant certain death in a few days. Since all of the victims lived for years after the incident, Sparks suggests the cause of the symptoms was some kind of chemical contamination, presumably by an aerosol.

Vickie Landrum telephoned a number of U.S. government agencies and officials about the encounter. When she telephoned NASA, Landrum was steered toward NASA aerospace engineer John Schuessler, long interested in UFOs. With some associates from civilian UFO research group MUFON, Schuessler began research on the case, and later wrote articles and a book on the subject. Astronomer Allan Hendry of CUFOS also briefly investigated the Cash-Landrum case.

Due to the Chinook helicopters' presence, the witnesses presumed that at least one branch of the United States Armed Forces witnessed the object, if they were not escorting or pursuing it. However, investigators could find no evidence linking the helicopters with any branch of the military.

In 1982, Lt. Col. George Sarran of the Department of the Army Inspector General began the only thorough formal governmental investigation into the supposed UFO encounter. He could not find any evidence that the helicopters the witnesses claimed to have seen belonged to the U.S. Armed Forces. Sarran stated that "Ms. Landrum and Ms. Cash were credible … the policeman and his wife [who claimed to have seen 12 helicopters near the UFO encounter site] were also credible witnesses. There was no perception that anyone was trying to exaggerate the truth." (quoted in Clark, 177)

In 1998, journalist and UFO sceptic Philip J. Klass, found a few reasons to doubt the story by Cash and Landrum:

> when Schuessler inspected Betty's car in early 1981 and used a geiger counter to check for radioactivity, he found none. Presumably [sic] he also checked for radioactivity when he visited the site of the (alleged) incident, and found no abnormal radiation ... [Schuessler] provides NO medical data on Betty's health PRIOR to the UFO incident. Nor does he provide any medical data on the prior health of Vicki or Colby. (emphasis in original)

Other UFO researchers point out that high-energy ionizing radiation of the kind that can cause damage to human beings (e.g. gamma radiation) does not induce radioactivity in objects, and would not have left behind any residual radioactivity in the area.

Similarly, Brookesmith writes, "Sceptics have always asked a blunt and fundamental question: what was the trio's state of health before their alleged encounter?"

In 1994, UFO skeptic Steuart Campbell suggested that the witnesses may have observed a mirage of Canopus, which lay exactly in line with the road.

01-08-1981 The **Trans-en-Provence Case** is one of the rare cases where an unidentified flying object is claimed to have left physical evidence, in the form of burnt residue from a field. The event took place on January 8, 1981, outside the town of Trans-en-Provence in the French *département* of Var. It was described in *Popular Mechanics* as "perhaps the most completely and carefully documented sighting of all time".

The case began on January 8, 1981 at 5pm. Renato Nicolaï, a fifty-five-year-old farmer, heard a strange whistling sound while performing agricultural work on his property. He then saw a saucer-shaped object about eight feet in diameter land about 50 yards (46 m) away at a lower elevation.

According to the witness, "The device had the shape of two saucers, one inverted on top of the other. It must have measured about 1.5 meters in height. It was the color of lead. This device had a ridge all the way around its circumference. Under the machine I saw two kinds of pieces as it was lifting off.

They could be reactors or feet. There were also two other circles which looked like trapdoors. The two reactors, or feet, extended about 20 cm below the body of the machine."

Nicolaï claimed the object took off almost immediately, rising above the treeline and departing to the north east. It left burn marks on the ground where it had sat.

The local gendarmerie were notified of the event the following day by Mr. Nicolaï directly on the advice of his neighbor's wife, Mrs Morin. The gendarmerie proceeded to interview Mr. Nicolaï, take photos of the scene, and collect soil and plant samples from the field. The case was later sent to GEIPAN—or GEPAN (*Groupe d'Étude des Phénomènes Aérospatiaux Non-identifiés*) as it was known at that time—for review.

GEPAN analysis noted that the ground had been compressed by a mechanical pressure of about 4 or 5 tons, and heated to between 300° and 600° C. Trace amounts of phosphate and zinc were found in the sample material, and analysis of resident alfalfa near the landing site showed chlorophyll levels between 30 and 50 percent lower than expected.

Mr. Nicolaï had initially believed the object to be an experimental military device. The close proximity of the site to the Canjuers military base makes such a theory generally plausible.

However, GEPAN's investigation focused on conventional explanations, such as atmospheric or terrain causes of a terrestrial nature. But despite a joint investigation by GEPAN and the gendarmerie which lasted for two years no plausible explanation was found.

12-31-1981 Hudson Valley Sightings/Hudson Valley Boomerang. A wave of reported UFO sightings in the Hudson Valley which ran until 1987, peaking in 1983-84.

06-30-1983 Copely Woods Encounter. Hundreds of Basketball-Sized ball of lights were sighted around a neighborhood, leaving unusually obvious marks behind. Budd Hopkins wrote his book "Intruders" about the case.

12-27-1985 Communion. Strieber asserts that he was abducted from his cabin in upstate New York on the evening of December 26, 1985 by non-human beings. He wrote about this experience and related experiences in *Communion* (1987), his first non-fiction book. Although the book is perceived generally as an account of alien abduction, Strieber draws no conclusions about the identity of alleged abductors. He refers to the beings as "the visitors," a name chosen to be as neutral as possible to entertain the possibility that they are not extraterrestrials and may instead exist in his mind.

Both the hardcover and paperback edition of *Communion* reached the number one position on the New York Times Best Seller list (non-fiction), with more than 2 million copies collectively sold. With *Communion*, an esoteric subject had reached the cultural mainstream, and Strieber found himself, perhaps unexpectedly, as its representative.

Following the popularity of the book, the author's account was subject to intense scrutiny and even derision. Some disparagement came from within the publishing world itself:

Although published as non-fiction, the book editor of the *Los Angeles Times* pronounced the follow-up title, *Transformation* (1988), to be fiction and removed it from the non-fiction best-seller list (it nonetheless made the top 10 on the fiction side of the chart). "It's a reprehensible thing," Strieber responded. "My book is a true story ... Placing this book on the fiction list is an ugly example of exactly the kind of blind prejudice that has hurt human progress for many generations." Criticism noting the similarity between the non-human beings in Strieber's autobiographical accounts and the non-human beings in his initial horror novels were typically acknowledged by the author as a fair observation, but not indicative of his autobiographical works being fictional: "The mysterious small beings that figure prominently in *Catmagic* seem to be an unconscious rendering of [the visitors], created before I was aware that they may be real."

Over the next 24 years (since the 1987 publication of *Communion*), Strieber wrote four additional autobiographies detailing his experiences with the visitors: *Transformation* (1988), a direct follow-up; *Breakthrough: The Next Step* (1995), a reflection on the original events and accounts of the sporadic contact he'd subsequently experienced; *The Secret School* (1996), in which he examines strange memories from his childhood; and lastly, *Solving the Communion Enigma: What Is to Come* (2011).

In *Solving the Communion Enigma*, Strieber reflects on how advances in scientific understanding since his 1987 publication may shed light on what he perceived, noting, "Among other things, since I wrote *Communion*, science has determined that parallel universes may be physically real and that time travel may in some way be possible".

The book is a consolidation of UFO sightings and related phenomena, including crop circles, alien abductions, mutilations and deaths in an attempt to discern any kind of meaningful overall pattern. Strieber concludes that the human species is being shepherded to a higher level of understanding and beingness within an endless "multiverse" of matter, energy, space and time. He also writes more candidly about the deleterious effects his initial experiences had upon him while staying at his upstate New York cabin in the 1980s, noting "I was regularly drinking myself to sleep when we were there. I would listen to the radio until late hours, drinking vodka..."

Other visitor-themed books of Strieber's include *Majestic* (1989), a novel about the Roswell UFO incident; *The Communion Letters* (1997, reissued in 2003), a collection of letters from readers reporting experiences similar to Strieber's; *Confirmation* (1998), in which Strieber reviews a variety of evidence that is suggestive of alien contact, and considers what more would be required to provide 'confirmation'; *The Grays* (2006) a novel in which his impressions of alien contact are presented through a fictional thriller/espionage narrative, and; *Hybrids* (2011) a fictional narrative that imagines human/alien hybrids being born into the modern world.

Additional visitor-themed writings include a screenplay for the 1989 film *Communion*, directed by Philippe Mora and starring Christopher Walken as Strieber. The movie covers material from the novel *Communion* and *Transformation*. Strieber has stated that he was dissatisfied with the film, which utilized scenes of improvised dialogue and includes themes not present in his books. Strieber also wrote a screenplay for his novel *Majestic*, which has not been filmed as of 2015.

Whitley Strieber has repeatedly expressed frustration that his experiences have been taken as "alien contact" when he does not actually know what they were. Strieber has reported anomalous childhood experiences and suggested that he may have suffered some sort of early interference by intelligence or military agencies.

He was extensively tested for temporal lobe epilepsy and other brain abnormalities at his own request, but his brain was found to be functioning normally. The results of these tests were reported in his book *Transformation*.

In the predawn hours of June 6, 1998, Strieber was allegedly visited in his Toronto hotel room by a mysterious but very ordinary-looking elderly Caucasian man, who delivered an unsolicited lecture covering various subjects from spirituality to the environment. When queried, the man airily suggested that he might be called "Michael," but Whitley has taken to referring to him as the "Master of the Key." Strieber first reported the visit in his online journal in 1998 and later gave a more complete account in his self-published book *The Key* (2001). Skeptics have pointed out that *The Key* and the 1998 journal entries give different (not contradictory but non-overlapping) accounts of what the man said. Strieber's mention of his personally devised system of shorthand or abbreviated note-taking in an interview with George Knapp on June 19, 2011, might at least partially account for this apparent discrepancy as the author had to reconstruct the entire 45-minute conversation with his visitor from a series of barely legible squiggles he discovered by his hotel bedside upon reawakening from deep sleep much later that same morning. He also chose to emphasize different subjects or aspects of the exchange according to how he surmised they could best be assimilated by his readers.

Strieber claims that the stranger in his room informed him that humans have an electron floating in front of their foreheads, and that that may indeed be their soul. He also claimed the stranger handed him a vial of unknown white liquid, instructed him to drink it, and he did.

Before publishing *The Key*, Strieber coauthored, with Art Bell, *The Coming Global Superstorm* (1999), a book about the possibility of rapid and destructive climate change. He has said that it was based largely on things the Master of the Key had told him about the environment. The book served as the inspiration for the disaster film *The Day After Tomorrow* (2004) and Strieber later wrote a novelization of that movie.

Another book Strieber says was inspired by the teachings of the Master of the Key is the self-published *The Path* (2002), which deals with the symbolism of the Tarot of Marseilles.

05-19-1986 São Paulo UFO sighting. Brazilian Air Force allegedly detect and intercept UFOs in southeastern Brazil.

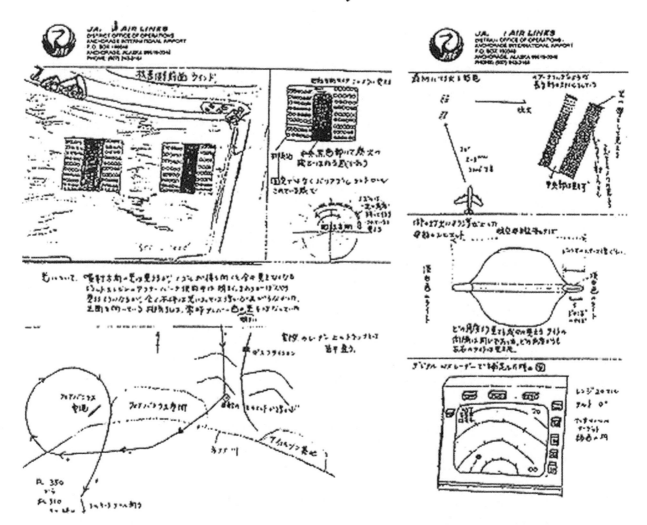

Japan Air Lines flight 1628 was a UFO incident that occurred on November 17, 1986 involving a Japanese Boeing 747 cargo aircraft. The aircraft was en route from Paris to Narita, Tokyo with a cargo of Beaujolais wine. On the Reykjavík to Anchorage section of the flight, at 5:11 PM over eastern Alaska, the crew first witnessed two unidentified objects to their left. These abruptly rose from below and closed in to escort their aircraft. Each had two rectangular arrays of what appeared to be glowing nozzles or thrusters, though their bodies remained obscured by darkness. When closest, the aircraft's cabin was lit up and the captain could feel their heat in his face.

These two craft departed before a third, much larger disk-shaped object started trailing them, causing the pilots to request a change of course. Anchorage Air Traffic Control obliged and requested an oncoming United Airlines flight to confirm the unidentified traffic, but when it and a military craft sighted JAL 1628 at about 5:51 PM, no other craft could be distinguished. The sighting of 50 minutes ended in the vicinity of Denali.

On November 17, 1986, the Japanese crew of a JAL Boeing 747 cargo freighter witnessed three unidentified objects after sunset, while flying over eastern Alaska, USA. The objects seemed to prefer the cover of darkness to their left, and to avoid the brighter skies to their right. At least the first two of the objects were observed by all three crew members: Captain Kenju Terauchi, an ex-fighter pilot with more than 10,000 hours flight experience, in the cockpit's left-hand seat; co-pilot Takanori Tamefuji in the right-hand seat; and flight engineer Yoshio Tsukuba.

The routine cargo flight entered Alaska on auto-pilot, cruising at 565 mph (909 km/h) at an altitude of 35,000 ft (11,000 m). At 5:09 PM, the Anchorage ATC advised a new heading towards Talkeetna, Alaska.

As soon as JAL 1628 straightened out of its turn, at 05:11 PM, Captain Terauchi noticed two craft to his far left, and some 2,000 ft (610 m) below his altitude, which he assumed to be military aircraft. These were pacing his flight path and speed. At 5:18 or 5:19 PM the two objects abruptly veered to a position about 500 ft (150 m) or 1,000 ft (300 m) in front of the aircraft, assuming a stacked configuration.

In doing so they activated "a kind of reverse thrust, and [their] lights became dazzlingly bright". To match the speed of the aircraft from their sideways approach, the objects displayed what Terauchi described as a disregard for inertia: "The thing was flying as if there was no such thing as gravity. It sped up, then stopped, then flew at our speed, in our direction, so that to us it [appeared to be] standing still. The next instant it changed course. ... In other words, the flying object had overcome gravity." The "reverse thrust" caused a bright flare for 3 to 7 seconds, to the extent that captain Terauchi could feel the warmth of their glows.

Air traffic control was notified at this point (i.e. 5:19:15 PM), who could not confirm any traffic in the indicated position. After 3 to 5 minutes the objects assumed a side-to-side configuration, which they maintained for another 10 minutes. They accompanied the aircraft with an undulating motion, and some back and forth rotation of the jet nozzles, which seemed to be under automatic control, causing them to flare with brighter or duller luminosity.

Each object had a square shape, consisting of two rectangular arrays of what appeared to be glowing nozzles or thrusters, separated by a dark central section. Captain Terauchi speculated in his drawings, that the objects would appear cylindrical if viewed from another angle, and that the observed movement of the nozzles could be ascribed to the cylinders' rotation. The objects left abruptly at about 5:23:13 PM, moving to a point below the horizon to the east.

Where the first objects disappeared, Captain Terauchi now noticed a pale band of light that mirrored their altitude, speed and direction.

Setting their onboard radar scope to a 25 nautical miles (46 km) range, he confirmed an object in the expected 10 o'clock direction at about 7.5 nmi (13.9 km) distance, and informed ATC of its presence. Anchorage found nothing on their radar, but Elmendorf ROCC, directly in his flight path, reported a "surge primary return" after some minutes.

As the city lights of Fairbanks began to illuminate the object, captain Terauchi believed to perceive the outline of a gigantic spaceship on his port side that was "twice the size of an aircraft carrier". It was however outside first officer Tamefuji's field of view. Terauchi immediately requested a change of course to avoid it. The object however followed him "in formation", or in the same relative position throughout the 45 degree turn, a descent from 35,000 to 31,000 ft, and a 360 degree turn. The short-range radar at Fairbanks airport however failed to register the object.

Anchorage ATC offered military intervention, which was declined by the pilot, due to his knowledge of the Mantell incident. The object was not noted by any of two planes which approached JAL 1628 to confirm its presence, by which time JAL 1628 had also lost sight of it. JAL 1628 arrived safely in Anchorage at 18:20.

Captain Terauchi cited in the official Federal Aviation Administration report that the object was a UFO. In December 1986, Terauchi gave an interview to two Kyodo News journalists. JAL soon grounded him for talking to the press, and moved him to a desk job. He was only reinstated as a pilot years afterwards, and retired eventually in north Kanto, Japan.

Kyodo News contacted Paul Steucke, the FAA public information officer in Anchorage on December 24, and received confirmation of the incident, followed by UPI on the 29th. The FAA's Alaskan Region consulted John Callahan, the FAA Division Chief of the Accidents and Investigations branch, as they wanted to know what to tell the media about the UFO. John Callahan was unaware of any such incident, considering it a likely early flight of a stealth bomber, then in development. He asked the Alaskan Region to forward the relevant data to their technical center in Atlantic City, New Jersey, where he and his superior played back the radar data and tied it in with the voice tapes by videotaping the concurrent playbacks.

A day later at FAA headquarters they briefed Vice Admiral Donald D. Engen, who watched the whole video of over half an hour, and asked them not to talk to anybody until they were given the OK, and to prepare an encompassing presentation of the data for a group of government officials the next day. The meeting was attended by representatives of the FBI, CIA and President Reagan's Scientific Study Team, among others. Upon completion of the presentation, all present were told that the incident was secret and that their meeting "never took place". According to Callahan, the officials considered the data to represent the first instance of recorded radar data on a UFO, and they took possession of all the presented data. John Callahan however managed to retain the original video, the pilot's report and the FAA's first report in his office. The forgotten target print-outs of the computer data were also rediscovered, from which all targets can be reproduced that were in the sky at the time.

After a three-month investigation, the FAA formally released their results at a press conference held on March 5, 1987. Here Paul Steucke retracted earlier FAA suggestions that their controllers confirmed a UFO, and ascribed it to a "split radar image" which appeared with unfortunate timing. He clarified that "the FAA [did] not have enough material to confirm that something was there", and though they were "accepting the descriptions by the crew" they were "unable to support what they saw". The McGrath incident was revealed here amongst the ample set of documents supplied to the journalists.

The sighting received special attention from the media, as a supposed instance of the tracking of UFOs on both ground and airborne radar, while being observed by experienced airline pilots, with subsequent confirmation by an FAA Division Chief.

On 29 January 1987 at 18:40 PM, Alaska Airlines Flight 53 observed a fast moving object on their onboard weather radar. While at 35,000 ft (11,000 m), some 60 miles (97 km) west of McGrath, on a flight from Nome to Anchorage, the radar registered a strong target in their 12 o'clock position, at 25 miles (40 km) range.

While they could not distinguish any object or light visually, they noticed that the radar object was increasing its distance at a very high rate. With every sweep of their radar, about 1 second apart, the object added five miles to its distance, translating to a speed of 18,000 mph (29,000 km/h). The pilot however relayed a speed of 'a mile a second' to the control tower, or a speed of 3,600 mph (5,800 km/h), but confirmed that the target exceeded both the 50 mi (80 km) and 100 mi (160 km) ranges of their radar scope in a matter of seconds.

The object was outside the radar range of the Anchorage ARTCC, and additional radar data covering the specified time and location failed to substantiate the pilots' claim.

A US Air Force KC-135 jet flying from Anchorage to Fairbanks once again observed a very large, disk-shaped object on January 30, 1987. The pilot reported that the object was 12 m (40 ft) from the aircraft. The object then disappeared out of sight.

11-11-1987 Gulf Breeze UFO incident. The **Gulf Breeze UFO incident** was a series of claimed UFO sightings in Gulf Breeze Florida during the late 1980s. The claims gained press and television coverage at the time, but are now widely considered a hoax.

Beginning on November 11, 1987 Gulf Breeze contractor Ed Walters reported a series of UFO sightings over a period of three weeks. Some ufologists believed the photographs that Walters purported to show a flying saucer were genuine, however others strongly suspected them to be a hoax. Press coverage given Walters claims during that time has been criticized as "uncritical" and "sensationalist". The hoax allegations were confirmed some years later when investigators found a model UFO in the house Walters had occupied.

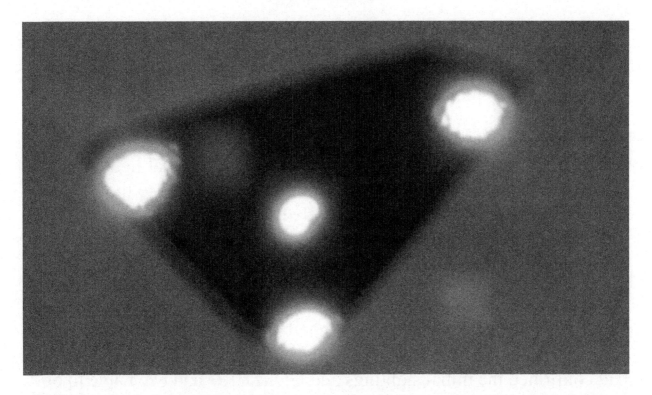

03-30-1990 The Belgian UFO wave. The Belgian UFO wave began in November 1989. The events of 29 November would be documented by no less than thirty different groups of witnesses, and three separate groups of police officers. All of the reports related a large object flying at low altitude. The craft was of a flat, triangular shape, with lights underneath. This giant craft did not make a sound as it slowly moved across the landscape of Belgium. There was free sharing of information as the Belgian populace tracked this craft as it moved from the town of Liege to the border of the Netherlands and Germany.

The Belgian UFO wave peaked with the events of the night of 30/31 March 1990. On that night unknown objects were tracked on radar, chased by two Belgian Air Force F-16s, photographed, and were sighted by an estimated 13,500 people on the ground – 2,600 of whom filed written statements describing in detail what they had seen.

Following the incident the Belgian air force released a report detailing the events of that night.

At around 23:00 on 30 March the supervisor for the Control Reporting Center (CRC) at Glons received reports that three unusual lights were seen moving towards Thorembais-Gembloux, which lies to the South-East of Brussels. The lights were reported to be brighter than stars, changing color between red, green and yellow, and appeared to be fixed at the vertices of an equilateral triangle. At this point Glons CRC requested the Wavre gendarmerie send a patrol to confirm the sighting.

Approximately 10 minutes later a second set of lights was sighted moving towards the first triangle. By around 23:30 the Wavre gendarmerie had confirmed the initial sightings and Glons CRC had been able to observe the phenomenon on radar. During this time the second set of lights, after some erratic manoeuvres, had also formed themselves into a smaller triangle. After tracking the targets and after receiving a second radar confirmation from the Traffic Center Control at Semmerzake, Glons CRC gave the order to scramble two F-16 fighters from Beauvechain Air Base shortly before midnight. Throughout this time the phenomenon was still clearly visible from the ground, with witnesses describing the whole formation as maintaining their relative positions while moving slowly across the sky. Witnesses also reported two dimmer lights towards the municipality of Eghezee displaying similar erratic movements to the second set of lights.

Over the next hour the two scrambled F-16s attempted nine separate interceptions of the targets. On three occasions they managed to obtain a radar lock for a few seconds but each time the targets changed position and speed so rapidly that the lock was broken.

During the first radar lock, the target accelerated from 240 km/h to over 1,770 km/h while changing altitude from 2,700 m to 1,500 m, then up to 3,350 m before descending to almost ground level – the first descent of more than 900 m taking less than two seconds. Similar manoeuvres were observed during both subsequent radar locks. On no occasion were the F-16 pilots able to make visual contact with the targets and at no point, despite the speeds involved, was there any indication of a sonic boom. Moreover, narrator Robert Stack added in an episode of *Unsolved Mysteries*, the sudden changes in acceleration and deceleration would have been fatal to one or more human pilots.

During this time, ground witnesses broadly corroborate the information obtained by radar. They described seeing the smaller triangle completely disappear from sight at one point, while the larger triangle moved up-wards very rapidly as the F-16s flew past. After 00:30 radar contact became much more sporadic and the final confirmed lock took place at 00:40. This final lock was once again broken by an acceleration from around 160 km/h to 1,120 km/h after which the radar of the F-16s and those at Glons and Semmerzake all lost contact. Following several fur-ther unconfirmed contacts the F-16s eventually returned to base shortly after 01:00.

The final details of the sighting were provided by the members of the Wavre gendarmerie who had been sent to confirm the original report. They describe four lights now being arranged in a square formation, all making short jerky movements, before gradually losing their luminosity and disappearing in four separate directions at around 01:30.

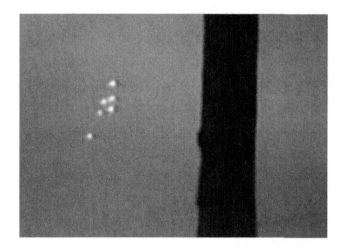

08-24-1990 Greifswald Lights. At 8,30 p.m. many people saw a group of 7 luminous spheres in the sky. The spheres took the form of a Y and were witnessed for about 30 minutes. Photographic evidence exists.

04-07-1991 Baviaanspoort Sighting. A hovering triangular craft with red central light, and white star-like lights on each extremity, was observed by a family at Baviaanspoort, Pretoria. A similar craft was sighted in the nearby Eersterust township on the evenings of 8 and 9 April, either stationary or moving.

09-15-1991 STS-48 incident. **STS-48** was a Space Shuttle mission that launched on 12 September 1991, from Kennedy Space Center, Florida. The orbiter was Space Shuttle *Discovery*. The primary payload was the Upper Atmosphere Research Satellite.

Video taken during mission STS-48 shows a flash of light and several objects, apparently flying in an artificial or controlled fashion. NASA explained them as ice particles reacting to engine jets.

08-08-1993 Kelly Cahill Abduction. Housewife Kelly Cahill reported seeing a large craft and beings with glowing red eyes.

08-11-1993 Sasolburg Sighting. Residents of Sasolburg observed a craft arriving from the direction of Vereeniging. The craft departed in a flash in the direction of Parys, but returned some three minutes later. The craft, similar in appearance to a water droplet, was observed to change color and shape. With time individual lights were distinguished, and the body was determined to be cigar-shaped. While contained in a yellowish to orange glow, it emitted a downward blue light, before once again departing in a flash, upwards. A nearby town resident claimed to have found imprints of a small craft's landing gear some two months later.

09-1994 Warrenton Sighting. A farmer claimed to have made repeated observations of a noisy, nighttime craft travelling at great speeds, besides what he described as a 'mother ship'. The craft's noise was compared to the sound of a helicopter or Volkswagen Beetle engine. The farmer's general claims were supported by four independent observers.

09-16-1994 Ariel UFO incident. 62 children saw a landed round aircraft and a small man next to it near their school.

03-1995 South Africa UFO Flap. A UFO flap swept South Africa from late March to mid April 1995, which was widely reported in the media.

05-25-1995 America West Airlines Flight 564. A 300–400 foot long cigar-shaped UFO with rotating strobe light followed an America West Boeing 757.

07-28-1996 Erasmuskloof Sighting. A glowing disc was sighted by Sergeant Becker near the Adriaan Vlok police station, Erasmuskloof, Pretoria. The pulsating light contained a red triangle and emitted bright green tentacles, while a radar operator at Johannesburg International confirmed its presence. A chase ensued involving some 200 policemen and a police helicopter. The helicopter chase was given up at 10,000 ft near Bronkhorstspruit, when the object made a vertical ascent.

10-05-1996 Westendorff UFO sighting. Pilot observes a UFO emerge from a mother craft.

12-02-1996 STS-80 incidents. A video taken during mission STS-80 of the Space Shuttle Columbia was analyzed by Mark J. Carlotto. It included three unusual phenomena: two slow-moving circular objects; a strange rapidly moving burst of light near the Earth's surface; and a number of object traces near the shuttle. The first two may be shuttle debris and an unusual atmospheric phenomenon. An analysis of the object traces near the shuttle suggested they were not shuttle debris or meteors, though James Oberg deemed them to be nearby sunlit debris.

The **Phoenix Lights** (also identified as "**Lights over Phoenix**") was a UFO sighting which occurred in Phoenix, Arizona, and Sonora, Mexico on Thursday, March 13, 1997.

Lights of varying descriptions were reported by thousands of people between 19:30 and 22:30 MST, in a space of about 300 miles (480 km), from the Nevada line, through Phoenix, to the edge of Tucson.

There were allegedly two distinct events involved in the incident: a triangular formation of lights seen to pass over the state, and a series of stationary lights seen in the Phoenix area. The United States Air Force later identified the second group of lights as flares dropped by A-10 Warthog aircraft that were on training exercises at the Barry Goldwater Range in southwest Arizona.

Witnesses claim to have observed a huge V-shaped (several football field sized), coherently-moving dark UFO (stars would disappear behind the object and reappear as it passed by), producing no sound, and containing five spherical lights or possibly light-emitting engines. Fife Symington, the governor at the time, was one witness to this incident. As governor he ridiculed the idea of alien origin, but several years later he called the lights he saw "otherworldly" after admitting he saw a similar UFO.

At about 18:55 PST (19:55 MST), a man reported seeing a V-shaped object above Henderson, Nevada. He said it was about the "size of a 747", sounded like "rushing wind", and had six lights on its leading edge. The lights reportedly traversed northwest to the southeast.

An unidentified former police officer from Paulden, Arizona is claimed to have been the next person to report a sighting after leaving his house at about 20:15 MST. As he was driving north, he allegedly saw a cluster of reddish or orange lights in the sky, comprising four lights together and a fifth light trailing them. Each of the individual lights in the formation appeared to the witness to consist of two separate point sources of orange light. He returned home and through binoculars watched the lights until they disappeared south over the horizon.

Lights were also reportedly seen in the areas of Prescott and Prescott Valley. At approximately 20:17 MST, callers began reporting the object was definitely solid, because it blocked out much of the starry sky as it passed over.

John Kaiser was standing outside with his wife and sons in Prescott Valley when they noticed a cluster of lights to the west-northwest of their position. The lights formed a triangular pattern, but all of them appeared to be red, except the light at the nose of the object, which was distinctly white. The object, or objects, which had been observed for approximately 2 to 3 minutes with binoculars, then passed directly overhead the observers, they were seen to "Bank to the right", and they then disappeared in the night sky to the southeast of Prescott Valley. The altitude could not be determined, however it was fairly low and made no sound whatsoever.

At the town of Dewey, 10 miles (16 km) east of Prescott, Arizona, six people saw a large cluster of lights while driving northbound on Highway 69.

Tim Ley, his wife Bobbi, his son Hal, and his grandson Damien Turnidge first saw the lights when they were above Prescott Valley about 65 miles (100 km) away from them. At first they appeared to them as five separate and distinct lights in an arc-shape like they were on top of a balloon, but they soon realized the lights appeared to be moving towards them. Over the next ten or so minutes they appeared to be coming closer and the distance between the lights increased and they took on the shape of an upside down V. Eventually when the lights appeared to be a couple of miles away the witnesses could make out a shape that looked like a 60-degree carpenter's square with the five lights set into it, with one at the front and two on each side. Soon the object with the embedded lights appeared to be coming right down the street where they lived about 100 to 150 feet (30 to 45 meters) above them, traveling so slowly it appeared to hover and was silent. The object then seemed to pass over their heads and went through a V opening in the peaks of the mountain range towards Squaw Peak Mountain and toward the direction of Phoenix Sky Harbor International Airport.

Witnesses in Glendale, a suburb northwest of Phoenix, saw the object pass overhead at an altitude high enough to become obscured by the thin clouds; this was at approximately between 20:30 and 20:45 MST.

When the triangular formation entered the Phoenix area, Bill Greiner, a cement driver hauling a load down a mountain north of Phoenix, described the second group of lights: "I'll never be the same. Before this, if anybody had told me they saw a UFO, I would've said, 'Yeah and I believe in the Tooth Fairy.' Now I've got a whole new view and I may be just a dumb truck driver, but I've seen something that don't belong here." Greiner stated that the lights hovered over the area for more than two hours.

A report came from a young man in the Kingman area who stopped his car at a public phone to report the incident. "[The] young man, en route to Los Angeles, called from a phone booth to report having seen a large and bizarre cluster of stars moving slowly in the northern sky".

On April 21, 2008, lights were again reported over North Phoenix by local residents. According to witnesses, the lights formed a vertical line, then spread apart and made a diamond shape. The lights also formed a U-shape at one time. Tony Toporek video taped the lights. He was talking to neighbors at 8 p.m. when the lights appeared. He grabbed his camera to get the lights on video. A valley resident reported that shortly after the lights appeared, three jets were seen heading west in the direction of the lights. An official from Luke Air Force Base denied any United States Air Force activity in the area. On April 22, 2008, a resident of Phoenix told a newspaper that the lights were nothing more than his neighbor releasing helium balloons with flares attached. The following day a Phoenix resident who declined to be identified in news reports stated he had attached flares to helium balloons and released them from his back yard. However, no name or pictures of the reported hoaxter were ever released, nor was anyone cited, ticketed or charged-from the supposed releasing of flares over a residential area that at the time was enduring a record drought.

Imagery of the Phoenix Lights falls into two categories: images of the triangular formation seen prior to 22:00 MST in Prescott and Dewey, and images of the 22:00 MST Phoenix event. Almost all known images are of the second event. All known images were produced using a variety of commercially available camcorders and cameras.

There are few known images of the Prescott/Dewey lights. Television station KSAZ reported that an individual named Richard Curtis recorded a detailed video that purportedly showed the outline of a spacecraft, but that the video had been lost. The only other known video is of poor quality and shows a group of lights with no craft visible.

the "V", which appeared over northern Arizona and gradually traveled south over nearly the entire length of the state, eventually passing south of Tucson — was the apparently "wedge-shaped" object reported by then-Governor Symington and many others. This event started at about 20:15 MST over the Prescott area, and was seen south of Tucson by about 20:45 MST.

Proponents of two separate events propose that the first event still has no provable explanation, but that some evidence exists that the lights were in fact airplanes. According to an article by reporter Janet Gonzales that appeared in the *Phoenix New Times*, videotape of the v shape shows the lights moving as separate entities, not as a single object;

a phenomenon known as illusory contours can cause the human eye to see unconnected lines or dots as forming a single shape.

Mitch Stanley, an amateur astronomer, observed high altitude lights flying in formation using a Dobsonian telescope giving 43× magnification. After observing the lights, he told his mother, who was present at the time, that the lights were aircraft. According to Stanley, the lights were quite clearly individual airplanes; a companion who was with him recalled asking Stanley at the time what the lights were, and he said, "Planes". When Stanley first gave an account of his observation at the Discovery Channel Town Hall Meeting with all the witnesses there he was shouted down in his assertion that what he saw was what other witnesses saw. Some have claimed that Stanley was seeing the Maryland National Guard jets flying in formation during a routine training mission at the Barry M. Goldwater bombing range south of Phoenix. It is possible that the Phoenix Lights Vee is actually a group of planes based on the explanation of a similar sighting in South California.

During the Phoenix event, numerous still photographs and videotapes were made, distinctly showing a series of lights appearing at a regular interval, remaining illuminated for several moments and then going out. These images have been repeatedly aired by documentary television channels such as the Discovery Channel and the History Channel as part of their UFO documentary programming.

The most frequently seen sequence shows what appears to be an arc of lights appearing one by one, then going out one by one. UFO advocates claim that these images show that the lights were some form of "running light" or other aircraft illumination along the leading edge of a large craft — estimated to be as large as a mile (1.6 km) in diameter —

hovering over the city of Phoenix. Other similar sequences reportedly taken over a half hour period show differing numbers of lights in a V or arrowhead array. Thousands of witnesses throughout Arizona also reported a silent, mile wide V or boomerang shaped craft with varying numbers of huge orbs. A significant number of witnesses reported that the craft was silently gliding directly overhead at low altitude. The first-hand witnesses consistently reported that the lights appeared as "canisters of swimming light", while the underbelly of the craft was undulating "like looking through water". However, skeptics claim that the video is evidence that mountains not visible at night partially obstructed views from certain angles, thereby bolstering the claim that the lights were more distant than UFO advocates claim.

UFO advocate Jim Dilettoso claimed to have performed "spectral analysis" of photographs and video imagery that proved the lights could not have been produced by a man-made source. Dilettoso claimed to have used software called "Image Pro Plus" (exact version unknown) to determine the amount of red, green and blue in the various photographic and video images and construct histograms of the data, which were then compared to several photographs known to be of flares. Several sources have pointed out, however, that it is impossible to determine the spectral signature of a light source based solely on photographic or video imagery, as film and electronics inherently alter the spectral signature of a light source by shifting hue in the visible spectrum, and experts in spectroscopy have dismissed his claims as being scientifically invalid. Normal photographic equipment also eliminates light outside the visible spectrum — *e.g.*, infrared and ultraviolet — that would be necessary for a complete spectral analysis. The maker of "Image Pro Plus", Media Cybernetic, has stated that its software is incapable of performing spectroscopic analysis.

Cognitech, an independent video laboratory, superimposed video imagery taken of the Phoenix Lights onto video imagery it shot during daytime from the same location. In the composite image, the lights are seen to extinguish at the moment they reach the Estrella mountain range, which is visible in the daytime, but invisible in the footage shot at night. A broadcast by local Fox Broadcasting Company affiliate KSAZ-TV claimed to have performed a similar test that showed the lights were in front of the mountain range and suggested that the Cognitech data might have been altered. Dr. Paul Scowen, visiting professor of Astronomy at Arizona State University, performed a third analysis using daytime imagery overlaid with video shot of the lights and his findings were consistent with Cognitech. The *Phoenix New Times* subsequently reported the television station had simply overlaid two video tracks on a video editing machine without using a computer to match the zoom and scale of the two images.

The second event was the set of nine lights appearing to "hover" over the city of Phoenix at around 10 pm. The second event has been more thoroughly covered by the media, due in part to the numerous video images taken of the lights. This was also observed by numerous people who may have thought they were seeing the same lights as those reported earlier.

The U.S. Air Force explained the second event as slow-falling, long-burning LUU-2B/B illumination flares dropped by a flight of four A-10 Warthog aircraft on a training exercise at the Barry Goldwater Range at Luke Air Force Base. According to this explanation, the flares would have been visible in Phoenix and appeared to hover due to rising heat from the burning flares creating a "balloon" effect on their parachutes, which slowed the descent. The lights then appeared to wink out as they fell behind the Sierra Estrella, a mountain range to the southwest of Phoenix.

A Maryland Air National Guard pilot, Lt. Col. Ed Jones, responding to a March 2007 media query, confirmed that he had flown one of the aircraft in the formation that dropped flares on the night in question. The squadron to which he belonged was in fact at Davis-Monthan AFB, Arizona on a training exercise at the time and flew training sorties to the Barry Goldwater Range on the night in question, according to the Maryland Air National Guard. A history of the Maryland Air National Guard published in 2000 asserted that the squadron, the 104th Fighter Squadron, was responsible for the incident. The first reports that members of the Maryland Air National Guard were responsible for the incident were published in The Arizona Republic newspaper in July 1997.

Military flares such as these can be seen from hundreds of miles given ideal environmental conditions. Later comparisons with known military flare drops were reported on local television stations, showing similarities between the known military flare drops and the Phoenix Lights. An analysis of the luminosity of LUU-2B/B illumination flares, the type which would have been in use by A-10 aircraft at the time, determined that the luminosity of such flares at a range of approximately 50–70 miles would fall well within the range of the lights viewed from Phoenix. LUU-2 flares have a burn time of approximately 5 minutes when suspended from a parachute.

Dr Bruce Maccabee did an extensive triangulation of the four videotapes, determining that the objects were near or over the Goldwater Proving Grounds. Page 5 of Dr. Maccabee's analysis refers to Bill Hamilton and Tom King's sighting position at Steve Blonder's home. Blonder has worked with Dr. Maccabee to fully include his sighting position in the triangulation report. Maccabee has also refined three other sighting positions and lines of sight in 2012.

Shortly after the lights, Arizona Governor Fife Symington III held a press conference, stating that "they found who was responsible". He proceeded to make light of the situation by bringing his aide on stage dressed in an alien costume. (Dateline, NBC). But in March 2007, Symington said that he had witnessed one of the "crafts of unknown origin" during the 1997 event, although he did not go public with the information. In an interview with *The Daily Courier* in Prescott, Arizona, Symington said, "I'm a pilot and I know just about every machine that flies. It was bigger than anything that I've ever seen. It remains a great mystery. Other people saw it, responsible people. I don't know why people would ridicule it". Symington had earlier said, "It was enormous and inexplicable. Who knows where it came from? A lot of people saw it, and I saw it too. It was dramatic. And it couldn't have been flares because it was too symmetrical. It had a geometric outline, a constant shape."

Symington also noted that he requested information from the commander of Luke Air Force Base, the general of the National Guard, and the head of the Arizona Department of Public Safety. But none of the officials he contacted had an answer for what had happened, and were also perplexed. Later, he responded to an Air Force explanation that the lights were flares: "As a pilot and a former Air Force Officer, I can definitively say that this craft did not resemble any man made object I'd ever seen. And it was certainly not high-altitude flares because flares don't fly in formation". In an episode of the television show UFO Hunters called "The Arizona Lights", Symington said that he contacted the military asking what the lights were. The response was "no comment". He pointed out that he was the governor of Arizona at the time, not just some ordinary civilian.

Frances Barwood, the 1997 Phoenix city councilwoman who launched an investigation into the event, said that of the over 700 witnesses she interviewed, "The government never interviewed even one".

There was minimal news coverage at the time of the incident. In Phoenix, a small number of local news outlets noted the event, but it received little attention beyond that. But on June 18, 1997, *USA Today* ran a front-page story that brought national attention to the case. This was followed by news coverage on the ABC and NBC television networks. The case quickly caught the popular imagination and has since become a staple of UFO-related documentary television, including specials produced by the History Channel and the Discovery Channel.

07-1997 Trichardt Sighting. A hovering, cylindrical light or shiny cloud, was filmed by Andreas Mathios in the sky above the town of Trichardt, in the current western Mpumalanga province. Besides Mathios, it was independently observed by three other persons around 6:50am. The light suddenly dropped and rose again before disappearing.

12-27-1998 Graaff-Reinet sighting. The Laubscher family videotaped a group of roundish triangular craft passing over the town of Graaff-Reinet, at about 25,000 ft. These were changing color and sometimes circled one another, before being overtaken by a much larger, shiny, gold-colored craft. At this point all the objects departed to a cloud bank on the horizon.

Michael Ryan

21ˢᵗ century Sightings

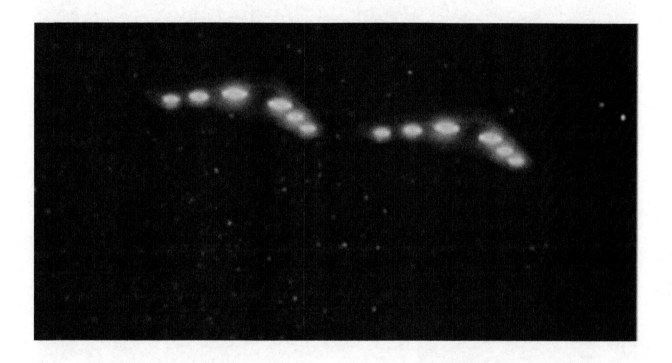

05-08-2000 Warden Sighting. Police inspector Kriel claimed to have observed an approaching UFO while travelling on the N3 freeway, 70 km north of Warden in the eastern Free State province. The orange, oval-shaped light was fitted with two cupolas, one above and another below, and was wide enough to cover four lanes of the freeway. After a close approach the craft receded again. A follow up report claimed that the vicinity is known for moving light apparitions.

The **Morristown UFO hoax** was originally thought to be an unidentified aerial event, of mysterious floating red lights, that occurred near
Morristown, New Jersey, on Monday, January 5, 2009, between 8:15 pm
and 9:00 pm. The lights were later observed on 4 other days: January 26,
January 29, February 7, and February 17, 2009. The events were actually
a hoax, meant as a social experiment.

Five flare lights attached to helium balloons were released by Joe Rudy
and Chris Russo and seen in the skies above Morris County, New Jersey.
Sightings were concentrated in the towns of Hanover Township,
Morristown, Morris Plains, Madison, and Florham Park.

On January 5, 2009, at 8:28 pm, the Hanover Township police department received the first of seven 9-1-1 calls. Neighboring police departments also received numerous phone calls with regard to the strange lights. Morristown Police Lt. Jim Cullen alerted Morristown Airport about a possible hazard to airplanes. Airport control tower workers reported seeing the lights in the sky, but could not determine what they were. Hanover Township police also contacted the Morristown Airport to try to pick up the objects on radar, but they were unable to pick up anything.

Major and local news networks covered the story, and Internet websites, including the Mutual UFO Network (MUFON), have posted information about the incident. On April 1, 2009, Rudy and Russo came forward with video evidence proving they were the perpetrators of this hoax, demonstrating how easy it is to fool the so-called UFO "experts."

On April 7, 2009, Russo and Rudy pleaded guilty to municipal charges of disorderly conduct and were sentenced to fines of $250 and 50 hours of community service.

On July 18, 2009, Russo and Rudy were the guest speakers for the New York City Skeptics at a public lecture in New York City, describing how they pulled off their hoax and the reasoning behind why they performed the hoax. On August 5, 2009, Russo was asked to debate a MUFON investigator on the existence of UFOs. The Morristown UFO Hoax was declared one of the best hoaxes ever on a program that aired on TruTV on April 1, 2015. Russo and Rudy were guests on the show.

On April 1, 2009, Russo and Rudy went public announcing that they had perpetrated this hoax to "show everyone how unreliable eyewitness accounts are, along with investigators of UFOs." As at least one police report suspected, Russo and Rudy had launched flares tied to helium balloons. Russo and Rudy described in detail how and why they perpetrated this hoax, and provided links to videos showing their preparations, the launch, and subsequent media coverage and involvement.

Two men from the Morristown area claimed to see the lights while driving on Hanover Avenue in Morris Plains. They recorded several videos and still photos of the event, which have been posted on news stations, websites, blogs, and YouTube. Rudy and Russo were interviewed on *News 12 New Jersey*, where they offered what would later be revealed to be a fictitious account of their sighting. They have since come forward as the perpetrators of the hoax resulting in the Morristown sightings. In the interview, Russo stated, "We were driving on Hanover, when all of a sudden we see these lights literally zip over our car." Rudy stated, "The lights seemed to ascend and descend almost in a sequence. They would rise up slowly and dip down."

A family in Hanover Township reported seeing the lights from their home. An 11-year-old, Kristin Hurley was the first to notice the lights. Paul Hurley, a pilot, saw the lights and said they were not planes. The Hurley Family took video of the lights, which appeared on Fox News. Hurley stated, "I have been in the aviation industry for 20 years and have never seen anything like this, a little scary, little scary."

A Morristown resident said that he saw an L-shaped formation oscillating in the sky. Bender was interviewed by the Morris County newspaper *Daily Record*. Bender stated that, what he saw "didn't seem manmade" and, "No way this could have been weather balloons."

Hanover Township's health officer said that he saw the lights while walking his dog in Madison at 8:38 pm. In contrast to local police reports, Van Orden stated the lights did not appear to be flares because they did not leave trails. He also said that they sometimes appeared to move against the wind, "These things were moving fast, holding formation, and then moving in three different directions; I don't know what it was."

Chris Russo and Joe Rudy built up the media attention by repeating the hoax over various parts of Morris County on four more occasions after the January 5 incident. The subsequent hoaxes and sightings took place on January 26, January 29, February 7, and February 17.

The largest cluster of lights occurred on February 17. Nine red lights were reported to be traveling in formation. Shortly after that sighting, Capt. Jeff Paul, a spokesman for Morris County Prosecutor Robert A. Bianchi, said that federal authorities have expressed concern that the objects might be a threat to flights on their final approach to Newark Liberty International Airport. The Federal Aviation Administration advised Paul that they would issue an advisory to aircraft in the area. Paul said "numerous" 911 calls were received on the evening of February 17 in Morris Plains, Morristown, Morris Township, Hanover, Denville, Parsippany, Montville and the Morris County Communications Dispatch center. The lights appeared to be traveling north, he said, and air traffic controllers at Morristown Airport reported that they appeared to be at an altitude of about 2,500 feet (760 m).

Dorian Vicente, 46, of Parsippany, said the lights caused traffic to slow on Route 80 East in Denville at 8:40 p.m. as people watched them floating overhead. There were nine lights, she said, and they were scattered at first. Then she said they aligned in a straight line. That is when she and several other cars pulled to the side of the highway to try to capture the lights on video. "It was the weirdest thing," she said. Ray Vargas, a witness to the lights on February 17, believed he witnessed something extraordinary. When interviewed by the media he stated, "If it's a hoax, it's a real good hoax. There were no flares, no streaks … they were almost as if they were communicating with each other."

Officials with the Morris County prosecutor's office called the military and determined that no military flights were in the area, Paul said. The prosecutor's office also contacted the FAA, the Office of Homeland Security and Preparedness, and the New Jersey State Police Regional Operations Intelligence Center.

Prosecutor Robert Bianchi used what he called a "measured approach" and filed disorderly-person charges, rather than charges of indictable offenses. Bianchi criticized the defendants for wasting police resources, posing a fire threat, and posing an aviation threat. The defendants plea-bargained and received a sentence fine of $250 each and 50 hours of community service at the Hanover Recreation Commission.

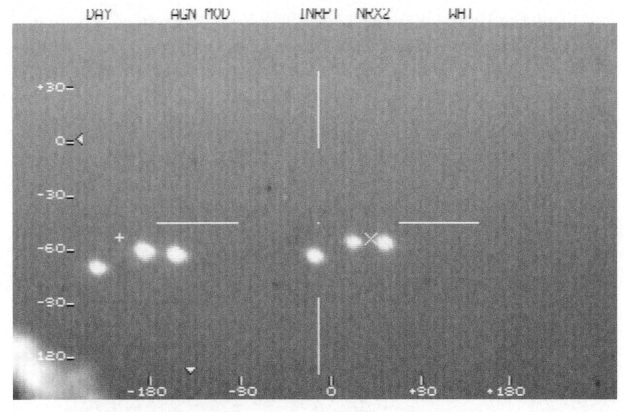

DAY   AGN MOD   INRPI NRX2   WHI

+30—

0=◁

-30—

-60—

-30—

-120—

-180   -90   0   +90   +180

LAT N 18°28.16' LON W 90°35.84'   -139.1°Az   2°El 05-03-04 17:06:49L

03-05-2004 2004 Mexican UFO incident. On Friday, March 5, Mexican Air Force pilots using infrared equipment to search for drug-smuggling aircraft recorded 11 unidentified objects over southern Campeche. Mexico's Defense Department issued a press release on May 12 accompanied by videotape that showed moving bright lights at 11,500 feet. Mexican UFOlogist Jaime Maussan interpreted the videotape as "proof of alien visitation", however science writer and skeptic Michael Shermer was critical of witness accounts that "varied wildly", saying "it was like a fisherman's tale, growing with each retelling", while other experts suggested the lights were most likely burn off flares from oil platforms.

08-21-2004 to 10-31-2006 The Tinley Park Lights. A sequence of five mass UFO sightings, first on August 21, 2004, two months later on October 31, 2004, again on October 1 of 2005, and once again on October 31, 2006, in Tinley Park and Oak Park, Chicago.

11-07-2006 2006 O'Hare International Airport UFO sighting. At approximately 16:15 CST on Tuesday, November 7, 2006, federal authorities at Chicago O'Hare International Airport received a report that a group of twelve airport employees were witnessing a metallic, saucer-shaped craft hovering over Gate C-17.

The object was first spotted by a ramp employee who was pushing back United Airlines Flight 446, which was departing Chicago for Charlotte, North Carolina. The employee apprised Flight 446's crew of the object above their aircraft. It is believed that both the pilot and co-pilot also witnessed the object.

Several independent witnesses outside of the airport also saw the object. One described a "blatant" disc-shaped craft hovering over the airport which was "obviously not clouds." According to this witness, nearby observers gasped as the object shot through the clouds at high velocity, leaving a clear blue hole in the cloud layer. The hole reportedly seemed to close itself shortly afterward.

According to the *Chicago Tribune's* Jon Hilkevitch, "The disc was visible for approximately two minutes and was seen by close to a dozen United Airlines employees, ranging from pilots to supervisors, who heard chatter on the radio and raced out to view it." So far, no photographic evidence of the UFO has surfaced, although it was reported to Hilkevitch that one of the United Airlines pilots was in possession of a digital camera at the time of the sighting and may have photographed the event.

Both United Airlines and the Federal Aviation Administration (FAA) first denied that they had any information on the O'Hare UFO sighting until the *Chicago Tribune*, which was investigating the report, filed a Freedom of Information Act (FOIA) request. The FAA then ordered an internal review of air-traffic communications tapes to comply with the *Tribune* FOIA request which subsequently uncovered a call by the United supervisor to an FAA manager in the airport tower concerning the UFO sighting.

The FAA stance concludes that the sighting was caused by a weather phenomenon and that the agency would not be investigating the incident. UFO investigators have pointed out that this stance is a direct contradiction to the FAA's mandate to investigate possible security breaches at American airports such as in this case;

an object witnessed by numerous airport employees and officially reported by at least one of them, hovering in plain sight, over one of the busiest airports in the world. Many witnesses interviewed by the *Tribune* were apparently "upset" that federal officials declined to further investigate the matter.

The Chicago O'Hare airport UFO story was picked up by various major mainstream media groups such as CNN, CBS, MSNBC, Fox News, *The Chicago Tribune*, and NPR.

On February 11, 2009, The History Channel aired an episode with the title *Aliens at the Airport* in which they reviewed the incident.

04-23-2007 **Alderney UFO sighting**. On April 23, 2007, captain **Ray Bowyer** was flying a routine passenger flight for the civilian airliner Aurigny Air Services, when he and his passengers gained progressively clearer views of two UFOs during a 12 to 15 minute period. Bowyer had 18 years of flying experience, and the 45-minute flight was one that he had completed every working day for more than 8 years.

Their 80 mi (130 km) journey of 45 minutes took them from Southampton on the southern coast of England, southwestwards to Alderney, being 10 miles (16 km) from France, and the northernmost of the Channel Islands. Their particular flight path had them converging on two enormous, seemingly stationary and identical airborne craft, which emanated brilliant yellow light.

A pilot of a plane near Sark, some 25 mi (40 km) to the south, confirmed the presence, general position and altitude of the first object from the opposite direction.

Radar traces also seemed to register the presence of an object, which Ray Bowyer believed to be correlated with the position and time of the sighting. A study by David Clarke however, could not establish a definite link, as the radar reflections of passenger ferries may have affected at least some of the readings.

10-19-2007 Kolkata UFO sighting. A fast moving object was spotted at 30° in the eastern horizon between 3:30AM and 6:30AM and filmed on handycam. Its shape shifted from a sphere to a triangle and then to a straight line. The object emitted a bright light forming a halo and radiated a range of colors. It was spotted by many people and hundreds gathered along the E.M. Bypass to catch a glimpse of the UFO, triggering a frenzy. The video footage was released on a TV News channel and later shown to Dr. D.P. Duari, the director of MP Birla Planetarium, Kolkata, who found it to be "extremely interesting and strange".

11-2007 to 12-2011 Dudley Dorito. The Dudley Dorito sightings concerns multiple sightings of a black triangle over the West Midlands conurbation of the United Kingdom which began in November 2007. The phrase was coined by the local press after hearing witness descriptions of the object.

01-08-2008 to 02-09-2009 Stephenville, Texas UFO sightings. UFOs were, and are sometimes still reported from this area. One was an object described as 1 mile (1.6 km) by 1.5 miles (2.4 km) in size, spotted over Bush Ranch in Crawford, Texas. The Air Force has identified the objects as training fighter jets that went unreported due to a "communications problem".

On January 8, 2008, Stephenville gained national media attention when dozens of residents reported observations of unidentified flying objects (UFOs).

Several residents described a craft as the size of a football field, while others said they were nearly a mile long, similar to the Phoenix lights mass sightings of March 13, 1997. Some observers reported military aircraft pursuing the objects.

CNN's Larry King covered the news story in the days following the incident, and according to Steve Allen, a private pilot who witnessed the UFO, the object was traveling at a high rate of speed which supposedly reached 3,000 feet in the air. Allen said it was "About a half a mile wide and about a mile long. It was humongous, whatever it was." The History Channel show UFO Hunters featured a story about the UFO sightings.

On January 23, after initially denying that any aircraft were operating in the area, the US Air Force said that it was conducting training flights in the Stephenville area that involved 10 fighter jets. The Air Force said they were merely F-16 Fighting Falcon jets conducting night flights from NAS JRB Fort Worth.

Washington Post blogger Emil Steiner reported that conspiracy theories had arisen claiming that reporter Angelia Joiner was fired from her job at the Empire-Tribune due to her reporting of the UFO story. Steiner added, "conjecture breeds conspiracy theories. Any official denial can be labeled a cover-up". Herald Tribune writer Billy Cox wrote that inquiries made about the incident on his personal UFO blog have been "stonewalled" by the USAF.

05-2008 to 09-2008 Turkey UFO sightings. Over a four-month span in 2008, a night guard at the Yeni Kent Compound videotaped one or more UFOs over Turkey at nighttime. Many witnesses confirmed the two and a half hours' worth of video, leading the Sirius UFO Space Science Research Center to dub it the "most important images of a UFO ever filmed".

06-20-2008 Wales UFO sightings. According to media reports, a police helicopter was almost hit by a UFO, before it tried to pursue it. Hundreds of people reported to have witnessed a UFO on the same or preceding days, from different areas of Wales.

02-27-2009 Middelburg Witbank Sighting. Two formations of high-flying, orange-red objects, were seen by many witnesses, and video-recorded by some, as they traveled between the towns of Middelburg and Witbank, 25 kilometers (16 mi) apart. The first formation of 7 objects were seen at 21:51 on 27 February, as they flew westwards from Middelburg towards Witbank. Due to their altitude they eventually disappeared behind clouds. At 20:00 on 6 March, they were noted again high in the sky, but this time greater in number, 23, and traveling in the opposite direction.

12-09-2009 The **Norwegian spiral anomaly of 2009** appeared in the night sky over Norway on 9 December 2009. It was visible from, and photographed from, northern Norway and Sweden. The spiral consisted of a blue beam of light with a greyish spiral emanating from one end of it. The light could be seen in all of Trøndelag to the south (the two red counties on the map to the right) and all across the three northern counties which compose Northern Norway, as well as from Northern Sweden and it lasted for 2–3 minutes According to sources, it looked like a blue light coming from behind a mountain, stopping in mid-air, and starting to spiral outwards. A similar, though less spectacular event had also occurred in Norway the month before. Both events had the expected visual features of failed flights of Russian SLBM RSM-56 Bulava missiles, and the Russian Defense Ministry acknowledged shortly after that such an event had taken place on 9 December.

01-25-2010 Harbour Mille incident. At least three UFOs were spotted over Harbour Mille. The objects looked like missiles but emitted no noise.

07-21-2010 Booysens Sighting. Residents of Booysens, Pretoria, observed a triangle of bright lights which hung motionless in the sky for two hours. In each instance the object commenced a slow descent towards the horizon at 20:30. Binocular observation revealed nothing more than a blue and emerald light, with a white light which shone straight downwards.

10-23-2010 Warren Air Force Base. UFO sighting coincides with 50 nuclear missiles going off-line.

01-28-2011 Jerusalem Dome of the Rock UFO incident. UFO seen hovering over Dome of the Rock. Then accelerating upward at high speed. This was later claimed to be a hoax.

02-20-2011 2011 Vancouver UFO sighting. Reports of purple and red flashing lights appearing in the sky on the evening of February 20, 2011 were revealed as a hoax by the owner of a large Chinese made kite with lights attached.

05-11-2011 Tierpoort Sighting. A host of silent, orange lights with consistent luminosity were observed as they traveled faster than a commercial aeroplane over Tierpoort near Pretoria (some 20 objects) and Krugersdorp respectively. On 15 June seven of these objects were observed and some photographed as they crossed the sky in single file over Tierpoort.

08-11-2014 Mass UFO sighting. Many people in Houston, Texas saw a ring of lights, flying during a thunderstorm and this was captured on video, object seemed to be transparent but the lights were arranged in a circle with a central light visible on some pictures.

10-03-2014 Colorado UFO sighting. The police in Breckenridge, Colorado received not only one but multiple reports of UFO sightings over the city. The mysterious objects remained stationary for periods of up to 15 minutes at a time. After that, a flash would appear and the "objects would take off across the mountain ridge".

11-11-2014 Airplane Disc Sighting. Plane passenger records speeding circular 'UFO' zooming beneath the jet he was in as it flies over Iran. Shaped like a disc, the fast-moving object does not resemble another plane, nor could it be mistaken for a helicopter.

06-25-2015 Kanpur UFO sighting. A schoolboy claimed to have captured photographs of a UFO.

Michael Ryan

Leaked Documents

Project Bluebook

DEPARTMENT OF THE AIR ~~RCE~~
WASHINGTON, D.C. 20330

OFFICE OF THE SECRETARY

FEDERAL GOVERNMENT

14 JAN 1977

Unidentified Flying Objects

Dear Mr. Malmfeldt:

This office recently received a public inquiry regarding UFOs, referred to us from the Bureau. As the inclosed fact sheet indicates, the Air Force's "Project Blue Book" investigation of UFOs was terminated on December 17, 1969, and all related documentation was turned over to the National Archives and Records Service.

The inquirer referred to us has been apprised of these events. Hopefully, the inclosed fact sheet will be of help in responding to any future inquiries on this subject.

Sincerely,

H. A. McCLANAHAN, Lt Col, USAF
Chief, Civil Branch
Community Relations Division
Office of Information

Attachment

Federal Bureau of Investigation
Attention: Mr. Malmfeldt, Room 7825
Washington, D. C. 20535

REC-26

62-83894-483

62-83894

AMERICAN REVOLUTION BICENTENNIAL
1776-1976

256

UFO FACT SHEET

On December 17, 1969 the Secretary of the Air Force announced the termination of Project Blue Book, the Air Force program for the investigation of UFOs.

The decision to discontinue UFO investigations was based on an evaluation of a report prepared by the University of Colorado entitled, "Scientific Study of Unidentified Flying Objects;" a review of the University of Colorado's report by the National Academy of Sciences; past UFO studies; and Air Force experience investigating UFO reports during the past two ·decades.

As a result of these investigations and studies, and experience gained from investigating UFO reports since 1948, the conclusions of Project Blue Book are: (1) no UFO reported, investigated, and evaluated by the Air Force has ever given any indication of threat to our national security; (2) there has been no evidence submitted to or discovered by the Air Force that sightings categorized as "unidentified" represent technological developments or principles beyond the range of present day scientific knowledge; and (3) there has been no evidence indicating that sightings categorized as "unidentified" are extraterrestrial vehicles.

With the termination of Project Blue Book, the Air Force regulation establishing and controlling the program for investigating and analyzing UFOs was rescinded. All documentation regarding the former Blue Book investigation has been permanently transferred to the Modern Military Branch, National Archives and Records Service, 8th and Pennsylvania Avenue, Washington, D.C. 20408, and is available for public review and analysis.

Attached for your information is the Project Blue Book sighting summary for the period 1947-1969. Also included is a listing of UFO-related materials currently available.

Since the termination of Project Blue Book, no evidence has been presented to indicate that further investigation of UFOs by the Air Force is warranted. In view of the considerable Air Force commitment of resources in the past, and the extreme pressure on Air Force funds at this time, there is no likelihood of renewed Air Force involvement in this area.

62-83894-483

TOTAL UFO SIGHTINGS, 1947 - 1969

YEAR	TOTAL SIGHTINGS	UNIDENTIFIED
1947	122	12
1948	156	7
1949	186	22
1950	210	27
1951	169	22
1952	1,501	303
1953	509	42
1954	487	46
1955	545	24
1956	670	14
1957	1,006	14
1958	627	10
1959	390	12
1960	557	14
1961	591	13
1962	474	15
1963	399	14
1964	562	19
1965	887	16
1966	1,112	32
1967	937	19
1968	375	3
1969	146	1
TOTAL	12,618	701

62-83894- 483

UFO MATERIALS

Scientific Study of Unidentified Flying Objects. Study
conducted by the University of Colorado under contract
F44620-76-C-0035. Three volumes, 1,465 p. 68 plates.
Photoduplicated hard copies of the official report may
be ordered for $6 per volume, $18 the set of three,
as AD 680:975, AD 680:976, and AD 680:977, from the
National Technical Information Service, U.S. Department
of Commerce, Springfield, VA 22151.

Review of University of Colorado Report on Unidentified
Flying Objects. Review of report by a panel of the National
Academy of Sciences. National Academy of Sciences,
1969, 6p. Photoduplicated hard copies may be ordered for
$3 as AD 688:541 from the National Technical Information
Service, U.S. Department of Commerce, Springfield, VA 22151.

There are a number of universities and professional
scientific organizations such as the American Association for
the Advancement of Science, which have considered UFO
phenomena during periodic meetings and seminars. In addition,
a list of private organizations interested in aerial phenomena
may be found in Gale's Encyclopedia of Associations (Edition 8,
Vol I, pp. 432-3). Such timely review of the situation by
private groups insures that sound evidence will not be overlooked by the scientific community.

2 Atchs
1. Sighting Summary
2. UFO-related Materials

2

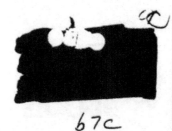

67c

SS *inspired*, and name-approved, by the United States Air Force.

67c

September 12, 1989

Mr. William S. Sessions
Director
Federal Bureau of Investigations
U.S. Department of Justice
Washington, D.C. 20535

Dear Mr. Sessions:

Your letter of September 1 has been received, and I am very
appreciative of hearing from you.

Your letter presented information that I had already mentioned
to you, in my letter of July 24; so this was information I knew.
I had given this to you as "advance" information - that you would
probably receive from other Government agencies, and departments,
and which is the standard release material.

Again, the new Project Blue Book is a civilian continuation of the
original Project Blue Book, which did close in 1969. Here, twenty-
years later, some of the same original Blue Book, Pentagon, and
other persons are, still, looking for the real answer to the UFO
phenomenon. That is the reason, too, I gave you the information
I did - about J. Edgar Hoover, President Carter, Senator Barry
Goldwater, etc. I was hoping to go "deeper" into the intelligence-
gathering agencies -which received some of the "better" UFO reports
which never did go to Project Blue Book.

I have been to the National Archives, and have looked at some of the
original Blue Book material that is there....I met with persons at
NASA's headquarters, in Washington, following their receipt of the
request from President Carter....I met with President Carter's chief
Scientific Advisor's personnel, at the New Executive Offices of the
President....I sent reports to the original Project Blue Book, and
know some of it's former personnel, etc. So, I know the whole scheme
of things, pertaining to UFO's - except for the hidden Intelligence-
gathering agency participation with regard to the better UFO sighting
reports, and which are, to this date, real, hard "unknowns".

Your response is appreciated and please accept my very best wishes.

Sincerely,

No ack
warranted.
See Bulet to
dated 9/1/89

67c

261

Inspired, and name-approved, by the *United States Air Force.*

July 24, 1989

Mr. William S. Sessions, Director
U..S. Department of Justice
Federal Bureau of Investigations
Washington, D.C. 20535

Dear Mr. Sessions:

You may recall that, while you were in Fort Smith visiting your
Father, I called and briefly mentioned my keen desire to meet
with you and to discuss the phenomenon of ████, in general,
and the role of U. S. Government participation, in particular.

I said that I wished to issue a "challenge" to you (not that you
do not challenges on a daily basis) pertaining to UFO's. You
stated that you were interested, and to write you - and which I
am now doing. I could write "pages" on the matter, but rather
than to do so - it will be lengthy enough for me to endeavor to
give you information I wish to convey in this letter.

In any event, the U.S. Government has played important roles
in the evaluation of UFO reported sightings. The new Project
Blue Book is a "continuation" of the original Project Blue Book,
established by the Air Force, and was located at Wright-Patterson
Air Force Base, Dayton, Ohio. Some of the most important sighting
"announcements" came from the Air Force, and which were, later,
stated to be something not in the original announcements - in a
cover-up of the initial announcement. This repeatedly happened
much to the chagrin of the Air Force. Original members of the
Project Blue Book found that many reports never were received by
this group, but went to intelligence-gathering agencies.

While Director of the F.B.I., Mr. J. Edgar Hoover (whose personnel
reported some of the sightings themselves and who were, otherwise,
brought in to some of the cases) repeatedly asked the Air Force
for complete information about the UFO's - to no avail. They (the
Air Force) would not release certain case reports to Mr. Hoover.
Even with the present F.O.I., mostly sanitized version/copies are
received and with information requested completely blocked-out.

Mr. William Ses   ns - F.B.I. - 2

While Chairman of the Senate Committee on Science, Senator Barry
Goldwater asked for information and was told that he was not on
a "need to know" basis. This request, of Goldwater, had to do with
an initial announcement from the Air Force that a "saucer" had crashed
and that alien bodies were (at that point in time) located in a hanger,
or other building, at Wright-Patterson Air Force Base, at Dayton, Ohio.
Senator Goldwater was refused any information in this matter, along wit
not being allowed to personally look inside of the alleged building.

Before taking office as President, Jimmy Carter announced to the public
that, if elected, he would give the American people the Government's
information about UFO's. This did not happen, and Mr. Carter was given
a complete run-around - and I could tell you more about this, because I
was contacted by the Pentagon on the matter.

The new Project Blue Book has 148 present, and/or former, Air Force, NC
NASA, and intelligence-gathering agency personnel, among others, acting
as advisor-consultants to PBB - anonymously. These are persons who, fo
the most part, have/had been actively engaged in earlier UFO reports fo
their respective agencies, and who are still muchly interested.

So, with this brief back-ground of information about me and about Gover
ment participation, the "challenge" I have mentioned - would be to have
you make inquiries at the highest levels of "officialdom" about what ou
Government really knows, whether or not there ARE alien bodies located
ANYWHERE, what happened to the crashed "flying saucers", where are the
photo's taken by Air Force personnel through on-board gun cameras, and
similar questioning. I am positive you would find that even you would
be getting a general run-around and could not receive satisfactory answ
to the above. I think you will/would find that you are not receiving
all the information you should be receiving and which should be very
evident to you after making several inquiries. You may tend to drop th
matter, thinking you are receiving answers to your questions, but this
will not be the case, I assure you.

Mr. William Sess. n; · F.B.I. - 3

Mr. Sessions, I will not take more of your time in this matter, but, if you are interested, I would appreciate your looking into the matter. You may wish to appoint one of your office personnel to start making inquiry about UFO 's, in general, and see which Agencies to which you would be directed. Keep in mind, however, that some present office personnel, in the other agencies, may not know what you would be inquiring about - and will tell you that the Air Force Project Blue Book closed it's "doors" in 1969, and that the Air Force no longer investigates UFO sighting reports - this I already know. The important key is to go beyond the A ir Force and into the other agencies; such as the Office of Defense, C.I.A., etc. President Bush, when asked about UFO's, told the person asking "You do not know the half of it". As you know, President Bush was formerly with the C.I.A. His comments are on tape, by the way.

Well, I think this should suffice for t his time. I appreciate your patience in reading this letter, and should you desire any additional information, please do not hesitate to let me know. I would like, very much, to work through you, and with you, to endeavor to bring the matte of UFO's to a conclusion - if that conclusion is an honest one.

I hope you had a nice visit,with your Dad, and c ome back for a visit again, soon.

Very sincerely,

67c

P.S. As sort of an "ultimate" challenge - why not ask President Bush, himself?

PROJECT
BLUE
BOOK

Inspired. and name-approved. by the
United States Air Force.

SEP 1 1989

SSP
CLASS
SRC'D   67c
SER
REC

67c

Dear

    I have received your July 24th letter in which you ask
for my help in securing information relative to investigations of
unidentified flying object (UFO) sightings.

    I have discussed your request with my colleagues, and I
would first like to explain that the investigation of UFOs is not
now nor has it ever been the responsibility of the FBI. The
Department of the Air Force conducted investigations and studies
of UFO reports from 1947 to 1969. On December 17, 1969, the
Secretary of the Air Force announced the termination of "Project
Blue Book," and the Air Force has furnished all documents
regarding its investigations of UFOs to the Modern Military
Branch, National Archives and Records Administration, Eighth
Street and Pennsylvania Avenue, N.W., Washington, D.C. 20408. I
understand this data is available for public review and analysis.
You may wish to make an inquiry of the Modern Military Branch for
the answers to some of your questions.

    Also of possible interest, the National Aeronautics and
Space Administration was asked by President Carter to look into
the possibility of resuming UFO investigations in 1977. After
studying all the facts available, it decided that nothing would
be gained by further investigation, and the Department of the
Air Force agreed with that decision. You may also want to check with
the National Aeronautics and Space Administration in connection
with your research.

63 - 0 - 92191

    I am not aware of any Federal agency tasked with the
responsibility of investigating UFOs and hope this information
will be of help to you and your organization.

                    Sincerely yours,

                    William S. Sessions

                    William S. Sessions
                    Director

63 - 0 - 92191

MAILED 21
SEP 8 1989

Exec AD Adm.
Exec AD Inv.
Exec AD LES
Asst. Dir.:
  Adm. Servs.
  Crim. Inv.
  Ident.
  Insp. _____ 1 - Little Rock - Enclosure
  Intell. _____ 1 - (Room 5042, TL 233)
  Lab. _____
  Legal Coun. _____ 1 -   67c
  Off. Cong. &
    Public Affs.
  Rec. Mgnt. _____ (6)   SEE NOTE PAGE TWO
  Tech. Servs.
  Training
Off. Liaison &
  Int. Affs.
Telephone Rm.
Director's Sec'y   MAIL ROOM ☐

244

67c

265

NOTE: ▓▓▓▓▓▓▓▓▓▓▓▓▓ who is not identifiable in Bufiles, **b7c.**
writes to the Director referencing what appears to be a brief
phone conversation he had with the Director while the Director,
was visiting his father in Fort Smith. He says he wants to
discuss UFOs and the role of U.S. Government participation in
particular, and he indicates the Director stated he was
interested and suggested he write to him ▓▓▓▓▓▓▓▓▓▓▓▓▓▓▓▓▓

▓▓▓▓▓▓▓▓ He asks the Director if he would make inquiries at
the highest levels of Government about what our Government really
knows about UFOs, whether or not there are alien bodies located
anywhere, what happened to the crashed "flying saucers," where
are the photos taken by Air Force personnel through on-board gun
cameras and similar questioning. Above response reviewed by
SSA ▓▓▓▓▓ Violent Crimes Unit, CID. Substance of paragraphs 2
and 3 of response previously provided to other inquiries on the
investigation of UFOs.

b7c

APPR		Adm Servs. _____	Off of Cong. _____
		Com. Inv _____	& Public Affs. _____
		Ident. _____	Off of Lia. _____
Director _____		Inspection _____	& Inst Affs. _____
Exec AD-Adm _____		Intell. _____	Rec Mgmt. _____
Exec AD-Inv _____		Laboratory _____	Tech Servs. _____
Exec AD-LES _____		Legal Coun. _____	Training _____

- 2 -

F.B.I. Letter

Office Memorandum • UNITED STATES GOVERNMENT

TO : DIRECTOR, FBI   DATE: August 12, 1947

FROM : SAC, MILWAUKEE

SUBJECT: FLYING DISCS
         SABOTAGE

        Reference is made to Bureau Bulletin No. 42, Series 1947, dated July 30, 1947, Section (B), which advises that all reports concerning flying discs should be investigated by field offices.

        Prior to the receipt of these instructions, two instances were called to the attention of this office concerning flying discs. One report was received July 7, 1947, the details of which are set forth in Milwaukee letter to the Bureau dated July 8, 1947, entitled, "Flying Discs or Saucers, Miscellaneous, Telephone Call from Mr. Fletcher at the Bureau at 8:30 a.m., 7-7-47." No investigation was conducted concerning this report.

        The second report was received by this office at 1:20 p.m. July 11, 1947, from ▓▓▓▓▓▓▓▓▓▓▓▓▓▓▓▓▓▓▓▓▓▓ who is in charge of the Civil Air Patrol of Wisconsin, an auxiliary of the Army Air Forces. On that occasion ▓▓▓▓▓▓▓▓▓▓▓▓ calling from Black River Falls, Wisconsin, telephonically advised this office that an object in the shape of a disc, nineteen inches in diameter had been found July 10, 1947, by one ▓▓▓▓▓▓▓ city electrician on the Jackson County fairgrounds, near Black River Falls, Wisconsin, about 3:30 p.m. The disc might be made of a substance such as cardboard covered by a silver airplane dope material. The contraption has a small wooden tail like a rudder in the back and inside of the disc is what appears to be an RCA photo-electric cell or tube. Also inside the disc is a little electric motor with a shaft running to the center of the disc. At one end of the shaft is a very small propeller. In ▓▓▓▓▓▓ opinion that contraption might possibly have been made by some juvenile. ▓▓▓▓▓▓▓▓▓▓ stated that he desired to return the contraption to Milwaukee and eventually turn it over to the Army Air Forces, but that the finder, ▓▓▓▓▓, apparently wanted to get some publicity on his find and wanted it returned to him.

        This information was telephonically called to the attention of Assistant Director D. M. LADD of the Bureau on July 11, 1947.

        Subsequently, SAC H. K. JOHNSON telephoned Colonel ▓▓▓▓▓▓ in charge of Counter Intelligence, Fifth Army, Chicago AC of S G-2 Headquarters Fifth Army, East Hyde Park Avenue, Chicago, Illinois, who stated he would contact ▓▓▓▓▓▓▓▓▓ of Black River Falls, Wisconsin.

        No further investigation was conducted in this matter.

COPIES DESTROYED
270 NOV 18 1964

James B. Shiley Report

HEADQUARTERS UNITED STATES AIR FORCE THE INSPECTOR GENERAL OFFICE OF SPECIAL INVESTIGATIONS REPORT OF INVESTIGATION	FILE NO. 24-34	DATE 14 March 1950

FILE NO. 24-34   DATE 14 March 1950

REPORT MADE BY
JAMES B. SHILEY

TITLE

RADIO IN POSSESSION OF MR. ~~█████~~
~~█████~~ WHICH ALLEGEDLY CAME FROM A FLYING
DISC THAT HAD CRASHED IN NEW MEXICO.

REPORT MADE AT
DO #18, Maywood, California
PERIOD 9,15,21 December 1949; 3,11,23,27,31
January; 17 February; 8 March 1950
OFFICE OF ORIGIN
DO #18, Maywood, California
STATUS
CLOSED

CHARACTER
UNCONVENTIONAL AIRCRAFT - SPECIAL IN~~████~~

CLASSIFICATION CANCELLED ~~████~~
BY AUTHORITY OF THE DIRECTOR OF SPEC IN.
RCRT A. KUNZE, Capt, USAF
BY Historian
5 DEC 1973
DATE

REFERENCE
None, this is an initial report.

SYNOPSIS

~~█████~~, motion picture actor, Beverly Hills, California, alleged
that one ~~█████~~ who claimed to be in the oil business, had stated
that a magnetic radio in his possession had come from a flying disc
which had crashed in New Mexico. ~~█████~~ apparently also stated that
the disc had contained men and that he had bits of cloth at his home
which came from clothing of these men, and that he also had pieces of
metal from gears of the disc. ~~█████~~ identified as being with the
~~█████~~ Denver, Colorado, and as residing at the ~~█████~~
~~█████~~ California. ~~█████~~ however, refused
to acknowledge requests left for him to communicate wi~~████~~ His
refusal may be due to fact that a local radio news commentator ridiculed
his story on a news broadcast.

UNCLASSIFIED

DISTRIBUTION		ACTION COPY FORWARDED TO	FILE STAMP
CG, AMC, Attn: MCIAXO-3, (Action Copies)	2	Commanding General Air Materiel Command	
Hq OSI	2	Wright-Patterson AFB	UNITED STATES AIR FORCE THE INSPECTOR GENERAL
DO #5	2	Dayton, Ohio	MAR 20 2 13 PM '50
File	2	ATTN: MCIA~~███~~	OFFICE OF SPECIAL INVESTIGATIONS

FOR OFFICIAL USE ONLY

KERIS O'K~~███~~
Lt. Col., USAF
DISTRICT COMMANDER

SMAMA—Nov 48—100M

Canada U.F.O. Document

DEFENCE RESEARCH BOARD

DRBS 200-4-160 (DSI)

ƒ 21-1-9

OTTAWA, Canada,
19 Apr 50.

D A I
Room 2621
"A" Building

Attention: W/C G S Austin

FLYING SAUCERS

1 Reference Minutes 19, 20 and 21 of the
220th JIC Meeting, please find attached draft *preliminary*
questionnaire for Service representatives and
RCMP.

2 I should be grateful if you would check
the questionnaire and add any questions to it
you may think fit.

3 We shall, presumably, have to produce a
number of copies of the questionnaire for distribution; perhaps you could indicate how many we
should run off to begin with.

4 I think we should also add to the questionnaire, an instruction as to whom it is to be sent
when completed. Would it be in order to say that
completed questionnaires are to be forwarded immediately to the Director of Air Intelligence,
National Defence Headquarters, Room 2621, "A"
Building, Ottawa?

5 You might also consider it advisable to
add that should the Flying Saucer actually make
a landing on Canadian territory, the nearest
RCAF Command should be advised immediately by
telegram or something of the sort.

6. Maybe we had better talk over this when
you have thought it over. A J G Langley —

Attach: 1

(A J G Langley)
Director of Scientific Intelligence

Classified Document

TOP SECRET/MJ-12

Central Intelligence Agency

From: Director of Central Intelligence (MJ-1)

To: MJ-2
MJ-3
MJ-4
MJ-5
MJ-6
MJ-7

Ref: Project MAJESTIC and JEHOVAH (MJ)
Project EVIRO
Project PARASITE
Project PARHELION

In the context of the above it has become necessary to review and evaluate duplication of field activities in light of the current situation. To eliminate this problem, I have drafted new directives for your review and consideration. Please evaluate each draft on its own merit with the goal of finding acceptable solutions in which all can agree on. As you must know LANCER has made some inquiries regarding our activities which we cannot allow. Please submit your views no later than October. Your action to this matter is critical to the continuance of the group.

Tab (A)   President's EYES ONLY
Tab (B)   "NEED-TO-KNOW"
Tab (C)   DoD 5200.1
Tab (D)   Project BLUE BOOK
Tab (E)   Freedom of Information
Tab (F)   PSYOP
Tab (G)   BW
Tab (H)   Project ENVIRONMENT

DO NOT REMOVE
FROM SAFE

Original carbon

TOP SECRET/MJ-12

U.F.O. Files

Michael Ryan

Top Secret Document

TOP SECRET / MAJIC
EYES ONLY
* TOP SECRET *

u 13

On 24 June, 1947, a civilian pilot flying over the Cascade Mountains in the State of Washington observed nine flying disc-shaped aircraft traveling in formation at a high rate of speed. Although this was not the first known sighting of such objects, it was the first to gain widespread attention in the public media. Hundreds of reports of sightings of similar objects followed. Many of these came from highly credible military and civilian sources. These reports resulted in independent efforts by several different elements of the military to ascertain the nature and purpose of these objects in the interests of national defense. A number of witnesses were interviewed and there were several unsuccessful attempts to utilize aircraft in efforts to pursue reported discs in flight. Public reaction bordered on near hysteria at times.

In spite of these efforts, little of substance was learned about the objects until a local rancher reported that one had crashed in a remote region of New Mexico located approximately seventy-five miles northwest of Roswell Army Air Base (now Walker Field).

On 07 July, 1947, a secret operation was begun to assure recovery of the wreckage of this object for scientific study. During the course of this operation, aerial reconnaissance discovered that four small human-like beings had apparently ejected from the craft at some point before it exploded. These had fallen to earth about two miles east of the wreckage site. All four were dead and badly decomposed due to action by predators and exposure to the elements during the approximately one week time period which had elapsed before their discovery. A special scientific team took charge of removing these bodies for study. (See Attachment "C".) The wreckage of the craft was also removed to several different locations. (See Attachment "B".) Civilian and military witnesses in the area were debriefed, and news reporters were given the effective cover story that the object had been a misguided weather research balloon.

* TOP SECRET *

TOP SECRET / MAJIC
EYES ONLY

TS0-EXEMPT (E)

004

278

Project Bluebook

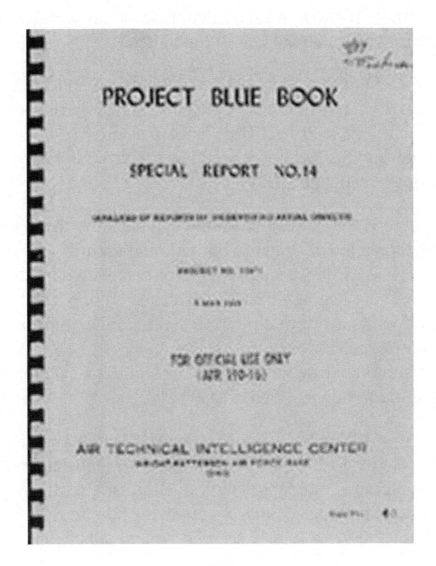

Project Blue Book was one of a series of systematic studies of unidentified flying objects (UFOs) conducted by the United States Air Force. It started in 1952, and it was the third study of its kind (the first two were projects Sign (1947) and Grudge (1949)). A termination order was given for the study in December 1969, and all activity under its auspices ceased in January 1970.

Project Blue Book had two goals:

1. To determine if UFOs were a threat to national security, and
2. To scientifically analyze UFO-related data.

Thousands of UFO reports were collected, analyzed and filed. As the result of the *Condon Report* (1968), which concluded there was nothing anomalous about UFOs, Project Blue Book was ordered shut down in December 1969 and the Air Force continues to provide the following summary of its investigations:

1. No UFO reported, investigated and evaluated by the Air Force was ever an indication of threat to our national security;
2. There was no evidence submitted to or discovered by the Air Force that sightings categorized as "unidentified" represented technological developments or principles beyond the range of modern scientific knowledge; and
3. There was no evidence indicating that sightings categorized as "unidentified" were extraterrestrial vehicles.

By the time Project Blue Book ended, it had collected 12,618 UFO reports, and concluded that most of them were misidentifications of natural phenomena (clouds, stars, etc.) or conventional aircraft. According to the National Reconnaissance Office a number of the reports could be explained by flights of the formerly secret reconnaissance planes U-2 and A-12.

A small percentage of UFO reports were classified as unexplained, even after stringent analysis. The UFO reports were archived and are available under the Freedom of Information Act, but names and other personal information of all witnesses have been redacted.

Public USAF UFO studies were first initiated under Project Sign at the end of 1947, following many widely publicized UFO reports (see Kenneth Arnold). Project Sign was initiated specifically at the request of General Nathan Twining, chief of the Air Force Materiel Command at Wright-Patterson Air Force Base. Wright-Patterson was also to be the home of Project Sign and all subsequent official USAF public investigations.

Sign was officially inconclusive regarding the cause of the sightings. However, according to US Air Force Captain Edward J. Ruppelt (the first director of Project Blue Book), Sign's initial intelligence estimate (the so-called Estimate of the Situation) written in the late summer of 1948, concluded that the flying saucers were real craft, were not made by either the Russians or US, and were likely extraterrestrial in origin. (See also extraterrestrial hypothesis.) This estimate was forwarded to the Pentagon, but subsequently ordered destroyed by Gen. Hoyt Vandenberg, USAF Chief of Staff, citing a lack of physical proof. Vandenberg subsequently dismantled Project Sign.

Project Sign was succeeded at the end of 1948 by Project Grudge, which was criticized as having a debunking mandate. Ruppelt referred to the era of Project Grudge as the "dark ages" of early USAF UFO investigation. Grudge concluded that all UFOs were natural phenomena or other misinterpretations, although it also stated that 23 percent of the reports could not be explained.

According to Captain Edward J. Ruppelt, by the end of 1951, several high-ranking, very influential USAF generals were so dissatisfied with the state of Air Force UFO investigations that they dismantled Project Grudge and replaced it with Project Blue Book in March 1952. One of these men was Gen. Charles P. Cabell. Another important change came when General William Garland joined Cabell's staff; Garland thought the UFO question deserved serious scrutiny because he had witnessed a UFO.

The new name, Project Blue Book, was selected to refer to the blue booklets used for testing at some colleges and universities. The name was inspired, said Ruppelt, by the close attention that high-ranking officers were giving the new project; it felt as if the study of UFOs was as important as a college final exam. Blue Book was also upgraded in status from Project Grudge, with the creation of the Aerial Phenomenon Branch.

Ruppelt was the first head of the project. He was an experienced airman, having been decorated for his efforts with the Army Air Corps during World War II, and having afterward earned an aeronautics degree. He officially coined the term "Unidentified Flying Object", to replace the many terms ("flying saucer" "flying disk" and so on) the military had previously used; Ruppelt thought that "unidentified flying object" was a more neutral and accurate term. Ruppelt resigned from the Air Force some years later, and wrote the book *The Report on Unidentified Flying Objects*, which described the study of UFOs by United States Air Force from 1947 to 1955. American scientist Michael D. Swords wrote that "Ruppelt would lead the last genuine effort to analyze UFOs".

Ruppelt implemented a number of changes: He streamlined the manner in which UFOs were reported to (and by) military officials, partly in hopes of alleviating the stigma and ridicule associated with UFO witnesses. Ruppelt also ordered the development of a standard questionnaire for UFO witnesses, hoping to uncover data which could be subject to statistical analysis. He commissioned the Battelle Memorial Institute to create the questionnaire and computerize the data. Using case reports and the computerized data, Battelle then conducted a massive scientific and statistical study of all Air Force UFO cases, completed in 1954 and known as "Project Blue Book Special Report No. 14".

Knowing that factionalism had harmed the progress of Project Sign, Ruppelt did his best to avoid the kinds of open-ended speculation that had led to Sign's personnel being split among advocates and critics of the extraterrestrial hypothesis. As Michael Hall writes, "Ruppelt not only took the job seriously but expected his staff to do so as well. If anyone under him either became too skeptical or too convinced of one particular theory, they soon found themselves off the project." In his book, Ruppelt reported that he fired three personnel very early in the project because they were either "too pro" or "too con" one hypothesis or another. Ruppelt sought the advice of many scientists and experts, and issued regular press releases (along with classified monthly reports for military intelligence).

Each U.S. Air Force Base had a Blue Book officer to collect UFO reports and forward them to Ruppelt. During most of Ruppelt's tenure, he and his team were authorized to interview any and all military personnel who witnessed UFOs, and were not required to follow the chain of command. This unprecedented authority underlined the seriousness of Blue Book's investigation.

Under Ruppelt's direction, Blue Book investigated a number of well-known UFO cases, including the so-called Lubbock Lights, and a widely publicized 1952 radar/visual case over Washington D.C.. According to Jacques Vallee, Ruppelt started the trend, largely followed by later Blue Book investigations, of not giving serious consideration to numerous reports of UFO landings and/or interaction with purported UFO occupants.

Astronomer Dr. J. Allen Hynek was the scientific consultant of the project, as he had been with Projects Sign and Grudge. He worked for the project up to its termination and initially created the categorization which has been extended and is known today as Close encounters. He was a pronounced skeptic when he started, but said that his feelings changed to a more wavering skepticism during the research, after encountering a minority of UFO reports he thought were unexplainable.

Ruppelt left Blue Book in February 1953 for a temporary reassignment. He returned a few months later to find his staff reduced from more than ten, to two subordinates. Frustrated, Ruppelt suggested that an Air Defense Command unit (the 4602nd Air Intelligence Service Squadron) be charged with UFO investigations.

In July 1952, after a build-up of hundreds of sightings over the previous few months, a series of radar detections coincident with visual sightings were observed near the National Airport in Washington, D.C. (see 1952 Washington D.C. UFO incident). Future Arizona Senator and 2008 presidential nominee John McCain is alleged to be one of these witnesses.

After much publicity, these sightings led the Central Intelligence Agency to establish a panel of scientists headed by Dr. H. P. Robertson, a physicist of the California Institute of Technology, which included various physicists, meteorologists, and engineers, and one astronomer (Hynek). The Robertson Panel first met on January 14, 1953 in order to formulate a response to the overwhelming public interest in UFOs.

Ruppelt, Hynek, and others presented the best evidence, including movie footage, that had been collected by Blue Book. After spending 12 hours reviewing 6 years of data, the Robertson Panel concluded that most UFO reports had prosaic explanations, and that all could be explained with further investigation, which they deemed not worth the effort.

In their final report, they stressed that low-grade, unverifiable UFO reports were overloading intelligence channels, with the risk of missing a genuine conventional threat to the U.S. Therefore, they recommended the Air Force de-emphasize the subject of UFOs and embark on a debunking campaign to lessen public interest. They suggested debunkery through the mass media, including Walt Disney Productions, and using psychologists, astronomers, and celebrities to ridicule the phenomenon and put forward prosaic explanations. Furthermore, civilian UFO groups "should be watched because of their potentially great influence on mass thinking… The apparent irresponsibility and the possible use of such groups for subversive purposes should be kept in mind."

It is the conclusion of many researchers that the Robertson Panel was recommending controlling public opinion through a program of official propaganda and spying. They also believe these recommendations helped shape Air Force policy regarding UFO study not only immediately afterward, but also into the present day.

There is evidence that the Panel's recommendations were being carried out at least two decades after its conclusions were issued (see the main article for details and citations).

In December 1953, Joint Army-Navy-Air Force Regulation number 146 made it a crime for military personnel to discuss classified UFO reports with unauthorized persons. Violators faced up to two years in prison and/or fines of up to $10,000.

In his book Ruppelt described the demoralization of the Blue Book staff and the stripping of their investigative duties following the Robertson Panel.

As an immediate consequence of the Robertson Panel recommendations, in February 1953, the Air Force issued Regulation 200-2, ordering air base officers to publicly discuss UFO incidents only if they were judged to have been solved, and to classify all the unsolved cases to keep them out of the public eye.

The same month, investigative duties started to be taken on by the newly formed 4602nd Air Intelligence Squadron (AISS) of the Air Defense Command. The 4602nd AISS was tasked with investigating only the most important UFO cases with intelligence or national security implications. These were deliberately siphoned away from Blue Book, leaving Blue Book to deal with the more trivial reports.

General Nathan Twining, who got Project Sign started back in 1947, was now Air Force Chief of Staff. In August 1954, he was to further codify the responsibilities of the 4602nd AISS by issuing an updated Air Force Regulation 200-2. In addition, UFOs (called "UFOBs") were defined as "any airborne object which by performance, aerodynamic characteristics, or unusual features, does not conform to any presently known aircraft or missile type, or which cannot be positively identified as a familiar object." Investigation of UFOs was stated to be for the purposes of national security and to ascertain "technical aspects." AFR 200-2 again stated that Blue Book could discuss UFO cases with the media only if they were regarded as having a conventional explanation. If they were unidentified, the media was to be told only that the situation was being analyzed. Blue Book was also ordered to reduce the number of unidentified to a minimum.

All this was done secretly. The public face of Blue Book continued to be the official Air Force investigation of UFOs, but the reality was it had essentially been reduced to doing very little serious investigation, and had become almost solely a public relations outfit with a debunking mandate. To cite one example, by the end of 1956, the number of cases listed as unsolved had dipped to barely 0.4 percent, from the 20 to 30% only a few years earlier.

Eventually, Ruppelt requested reassignment; at his departure in August 1953, his staff had been reduced from more than ten (precise numbers of personnel varied) to just two subordinates and himself. His temporary replacement was a noncommissioned officer. Most who succeeded him as Blue Book director exhibited either apathy or outright hostility to the subject of UFOs, or were hampered by a lack of funding and official support.

UFO investigators often regard Ruppelt's brief tenure at Blue Book as the high-water mark of public Air Force investigations of UFOs, when UFO investigations were treated seriously and had support at high levels. Thereafter, Project Blue Book descended into a new "Dark Ages" from which many UFO investigators argue it never emerged.

However, Ruppelt later came to embrace the Blue Book perspective that there was nothing extraordinary about UFOs; he even labeled the subject a "Space Age Myth."

In March 1954, Captain Charles Hardin was appointed the head of Blue Book. However, most UFO investigations were conducted by the 4602nd, and Hardin had no objection. Ruppelt wrote that Hardin "thinks that anyone who is even interested [in UFOs] is crazy. They bore him."

In 1955, the Air Force decided that the goal of Blue Book should be not to investigate UFO reports, but rather to reduce the number of unidentified UFO reports to a minimum. By late 1956, the number of unidentifed sightings had dropped from the 20-25% of the Ruppelt era, to less than 1%.

Captain George T. Gregory took over as Blue Book's director in 1956. Clark writes that Gregory led Blue Book "in an even firmer anti-UFO direction than the apathetic Hardin." The 4602nd was dissolved, and the 1066th Air Intelligence Service Squadron was charged with UFO investigations.

In fact, there was actually little or no investigation of UFO reports; a revised AFR 200-2 issued during Gregory's tenure emphasized that unexplained UFO reports must be reduced to a minimum.

One way that Gregory reduced the number of unexplained UFOs was by simple reclassification. "Possible cases" became "probable", and "probable" cases were upgraded to certainties. By this logic, a *possible* comet became a *probable* comet, while a probable comet was flatly declared to have been a misidentified comet. Similarly, if a witness reported an observation of an unusual balloon-*like* object,

Blue Book usually classified it as a balloon, with no research and quali-fication. These procedures became standard for most of Blue Book's lat-er investigations; see Hynek's comments below.

Major Robert J. Friend was appointed the head of Blue Book in 1958. Friend made some attempts to reverse the direction Blue Book had taken since 1954. Clark writes that "Friend's efforts to upgrade the files and catalog sightings according to various observed statistics were frustrated by a lack of funding and assistance."

Heartened by Friend's efforts, Hynek organized the first of several meet-ings between Blue Book staffers and ATIC personnel in 1959. Hynek suggested that some older UFO reports should be reevaluated, with the ostensible aim of moving them from the "unknown" to the "identified" category. Hynek's plans came to naught.

During Friend's tenure, ATIC contemplated passing oversight Blue Book to another Air Force agency, but neither the Air Research and De-velopment Center, nor the Office of Information for the Secretary of the Air Force was interested.

In 1960, there were U.S. Congressional hearings regarding UFOs. Civil-ian UFO research group NICAP had publicly charged Blue Book with covering up UFO evidence, and had also acquired a few allies in the U.S. Congress. Blue Book was investigated by the Congress and the CIA, with critics—most notably the civilian UFO group NICAP assert-ing that Blue Book was lacking as a scientific study. In response, ATIC added personnel (increasing the total personnel to three military person-nel, plus civilian secretaries) and increased Blue Book's budget. This seemed to mollify some of Blue Book's critics, that but it was only tem-porary.

A few years later (see below), the criticism would be even louder.

By the time he was transferred from Blue Book in 1963, Friend thought that Blue Book was effectively useless and ought to be dissolved, even if it caused an outcry amongst the public.

Major Hector Quintanilla took over as Blue Book's leader in August 1963. He largely continued the debunking efforts, and it was under his direction that Blue Book received some of its sharpest criticism. UFO researcher Jerome Clark goes so far as to write that, by this time, Blue Book had "lost all credibility."

Physicist and UFO researcher Dr. James E. McDonald once flatly declared that Quintanilla was "not competent" from either a scientific or an investigative perspective. However, McDonald also stressed that Quintanilla "shouldn't be held accountable for it", as he was chosen for his position by a superior officer, and was following orders in directing Blue Book.

Blue Book's explanations of UFO reports were not universally accepted, however, and critics — including some scientists — suggested that Project Blue Book was engaged in questionable research or, worse, perpetrating cover up. This criticism grew especially strong and widespread in the 1960s.

Take for example, the many mostly nighttime UFO reports from the midwestern and southeastern United States in the summer of 1965: Witnesses in Texas reported "multicolored lights" and large aerial objects shaped like eggs or diamonds. The Oklahoma Highway Patrol reported that Tinker Air Force Base (near Oklahoma City) had tracked up to four UFO's simultaneously, and that several of them had descended very rapidly: from about 22000 feet to about 4000 feet in just a few seconds, an action well beyond the capabilities of conventional aircraft of the era. John Shockley, a meteorologist from Wichita, Kansas, reported that, using the state Weather Bureau radar, he tracked a number of odd aerial objects flying at altitudes between about 6000 and 9000 feet. These and other reports received wide publicity.

Project Blue Book officially determined the witnesses had mistaken Jupiter or bright stars (such as Rigel or Betelgeuse) for something else.

Blue Book's explanation was widely criticized as inaccurate. Robert Riser, director of the Oklahoma Science and Art Foundation Planetarium offered a strongly worded rebuke of Project Blue Book that was widely circulated: "That is as far from the truth as you can get. These stars and planets are on the opposite side of the earth from Oklahoma City at this time of year. The Air Force must have had its star finder upside-down during August."

A newspaper editorial from the *Richmond News Leader* opined that "Attempts to dismiss the reported sightings under the rationale as exhibited by Project Bluebook (sic) won't solve the mystery … and serve only to heighten the suspicion that there's something out there that the air force doesn't want us to know about", while a Wichita-based UPI reporter noted that "Ordinary radar does not pick up planets and stars."

Another case that Blue Book's critics seized upon was the so-called Portage County UFO Chase, which began at about 5.00am, near Ravenna, Ohio on April 17, 1966. Police officers Dale Spaur and Wilbur Neff spotted what they described as a disc-shaped, silvery object with a bright light emanating from its underside, at about 1000 feet in altitude.

They began following the object (which they reported sometimes descended as low as 50 feet), and police from several other jurisdictions were involved in the pursuit. The chase ended about 30 minutes later near Freedom, Pennsylvania, some 85 miles away.

The UFO chase made national news, and the police submitted detailed reports to Blue Book. Five days later, following brief interviews with only one of the police officers (but none of the other ground witnesses), Blue Book's director, Major Hector Quintanilla, announced their conclusions: The police (one of them an Air Force gunner during the Korean War) had first chased a communications satellite, then the planet Venus.

This conclusion was widely derided, and was strenuously rejected by the police officers. In his dissenting conclusion, Hynek described Blue Book's conclusions as absurd: in their reports, several of the police had unknowingly described the moon, Venus *and* the UFO, though they unknowingly described Venus as a bright "star" very near the moon. Ohio Congressman William Stanton said that "The Air Force has suffered a great loss of prestige in this community … Once people entrusted with the public welfare no longer think the people can handle the truth, then the people, in return, will no longer trust the government."

In September 1968, Hynek received a letter from Colonel Raymond Sleeper of the Foreign Technology Division. Sleeper noted that Hynek had publicly accused Blue Book of shoddy science, and further asked Hynek to offer advice on how Blue Book could improve its scientific methodology. Hynek was to later declare that Sleeper's letter was "the first time in my 20 year association with the air force as scientific consultant that I had been officially asked for criticism and advice [regarding] … the UFO problem."

Quintanilla's own perspective on the project is documented in his manuscript, "UFOs, An Air Force Dilemma." Lt. Col Quintanilla wrote the manuscript in 1975, but it was not published until after his death. Quintanilla states in the text that he personally believed it arrogant to think human beings were the only intelligent life in the universe. Yet, while he found it highly likely that intelligent life existed beyond earth, he had no hard evidence of any extra terrestrial visitation.

In 1966, a string of UFO sightings in Massachusetts and New Hampshire provoked a Congressional Hearing by the House Committee on Armed Services. According to attachments to the hearing, the Air Force had at first stated that the sightings were the result of a training exercise happening in the area. But NICAP, the National Investigations Committee on Aerial Phenomena, reported that there was no record of a plane flying at the time the sightings occurred. Another report alleged that the UFO was actually a flying billboard promoting the gasoline. Raymond Fowler (of NICAP) added his own interviews with the locals, who saw Air Force officers confiscating newspapers with the story of UFOs and telling them not to report what they had seen. Two police officers who had witnessed the UFOs, Eugene Bertrand and David Hunt, wrote a letter to Major Quintanilla stating that they felt their reputations were destroyed by the Air Force. "It was impossible to mistake what we saw for any kind of military operation, regardless of altitude," the irritated officers wrote, adding that there was no way it could have been a balloon or helicopter. According to Secretary Harold Brown of the Air Force, Blue Book consisted of three steps: investigation, analysis, and the distribution of information gathered to interested parties. After Brown gave permission, the press were invited into the hearing. By the time of the hearing, Blue Book had identified and explained 95% of the reported UFO sightings. None of these were extraterrestrial or a threat to national security. Brown himself proclaimed, "I know of no one of scientific standing or executive standing with a detailed knowledge of this, in our organization who believes that they came from extraterrestrial sources." Dr. J. Allen Hynek, a science consultant to Blue Book, suggested in an unedited statement that a "civilian panel of physical and social scientists" be formed "for the express purpose of determining whether a major problem really exist" in regards to UFOs. Hynek remarked that he has "not seen any evidence to confirm" extraterrestrials, "nor do I know any competent scientist who has, or who believes that any kind of extraterrestrial intelligence is involved."

Criticism of Blue Book continued to grow through the mid-1960s. NICAP's membership ballooned to about 15,000, and the group charged the U.S. Government with a cover-up of UFO evidence.

Following U.S. Congressional hearings, the Condon Committee was established in 1966, ostensibly as a neutral scientific research body. However, the Committee became mired in controversy, with some members charging director Edward U. Condon with bias, and critics would question the validity and the scientific rigor of the Condon Report.

In the end, the Condon Committee suggested that there was nothing extraordinary about UFOs, and while it left a minority of cases unexplained, the report also argued that further research would not be likely to yield significant results.

In response to the Condon Committee's conclusions, Secretary of the Air Force Robert C. Seamans, Jr. announced that Blue Book would soon be closed, because further funding "cannot be justified either on the grounds of national security or in the interest of science."

The last publicly acknowledged day of Blue Book operations was December 17, 1969. However, researcher Brad Sparks, citing research from the May, 1970 issue of NICAP's *UFO Investigator*, reports that the last day of Blue Book activity was actually January 30, 1970. According to Sparks, Air Force officials wanted to keep the Air Force's reaction to the UFO problem from overlapping into a fourth decade, and thus altered the date of Blue Book's closure in official files.

Blue Book's files were sent to the Air Force Archives at Maxwell Air Force Base in Alabama. Major David Shea was to later claim that Maxwell was chosen because it was "accessible yet not too inviting."

Ultimately, Project Blue Book stated that UFOs sightings were generated as a result of:

- A mild form of mass hysteria.
- Individuals who fabricate such reports to perpetrate a hoax or seek publicity.
- Psychopathological persons.
- Misidentification of various conventional objects.

These official conclusions were directly contradicted by the USAF's own commissioned *Blue Book Special Report #14*. Psychological factors and hoaxes actually constituted less than 10% of all cases and 22% of all sightings, particularly the better-documented cases, remained unsolved. (See section below for details and Identified flying object.)

As of April 2003, the USAF has publicly indicated that there are no immediate plans to re-establish any official government UFO study programs.

Michael Ryan

UFO on the moon

The Third Kind

This transcripts was supposedly recorded during the Apollo 11 mission where it has been transcribed that the astronauts Neil Armstrong and Buzz Aldrin sighted UFO's on the moon.

During the broadcast of this historic event on the Canadian network coverage, they were discussing at some point a light which kept appearing while the astronauts were actually on the surface. Then it just seemed to be dropped.

One explanation for the halo's seen around or near some of the Appollo astronauts was that it was gases being vented from their backpacks.

Timothy Good writes that HAM radio operators receiving the VHF signals directly picked up the following message which was screened by NASA from the public.

Mission Control: What's there ? Mission Control calling Apollo 11.

Apollo 11: These babies are huge, sir ... enormous....Oh, God, you wouldn't believe it! I'm telling you there are other space craft out there... lined up on the far side of the crater edge... they're on the moon watching us.

Timothy Good uses "SAGA UFO SPECIAL #3" as a source for this quote.
From the book "Celestial Raise" by Richard Watson, ASSK, 1987, page 147-148;

"During the transmission of the Moon landing of Armstrong and Aldrin, who journeyed to the Moon in an American spaceship, two minutes of silence occurred in which the image and sound were interrupted. NASA insisted that this problem was the result of one of the television cameras which had overheated, thus interfering with the reception.

This unexpected problem surprised even the most qualified of viewers who were unable to explain how in such a costly project, one of the most essential elements could break down... Some time after the historic Moon landing, Christopher Craft, director of the base in Houston, made some surprising comments when he left NASA.

Author Sam Pepper (otherwise unidentified and he has since vanished) gave this version of "the top secret tape transcript" from "a leak close to the top", as follows:

Moon: Those are giant things. No, no, no - this is not an optical illusion. No one is going to believe this !

Huston: What ... what ... what ? What the h--- is happening ? What's wrong with you ?

Moon: They're here under the surface.

Huston: What's there ? (muffled noise) Emission interrupted; interference control calling 'Apollo 11'

Moon: We saw some visitors. They were here for a while, observing the instruments

The Third Kind

Huston: Repeat your last information !

Moon: I say that there were other spaceships. They're lined up in the other side of the crater !

Huston: Repeat, repeat !

Moon: Let us sound this orbita ... in 625 to 5 ... Automatic relay connected ... My hands are shaking so badly I can't do anything. Film it ? G--, if these d--ned cameras have picked up anything - what then ?

Huston: Have you picked up anything ?

Moon: I didn't have any film at hand. Three shots of the saucers or whatever they were that were ruining the film

Huston: Control, control here. Are you on your way ? What is the uproar with the UFOs over ?

Moon: They've landed here. There they are and they're watching us

Huston: The mirrors, the mirrors - have you set them up ?

Moon: Yes, they're in the right place. But whoever made those spaceships surely can come tomorrow and remove them. Over and out.

When the "Pepper Transcripts" first became public, UFO buffs wrote to their congressmen demanding that NASA officially confess to the coverup. NASA replied that "the incidents ... did not take place. Conversations between the Apollo 11 crew and Mission Control were released live during the entire Apollo 11 mission. There were between 1000 and 1500 representatives of the news media and T.V.

present at the Houston News Center listening and observing, and not one has suggested that NASA withheld any news or conversations of this nature." (Letter from Assistant Administrator for Legislative Affairs to several congressmen, January 1970.)

My Own UFO Experience

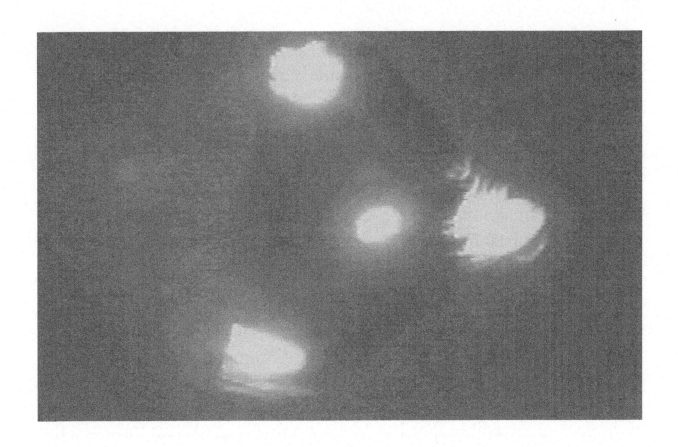

The late year of 2006 I saw this object in the sky. It was dusk but there was still plenty of light in the sky, bright enough that there was no shadowing of the objects around you. I was coming home from Manhattan, I had gotten off the train in Kings Park, New York and started to drive home. There was nothing unusual about the ride, it was very routine, I had taken the same way I had always taken to get home. Just before the turn into my block there is a row of close nit trees, after the turn my block starts a drop down a somewhat steep hill.

The trees before the turn where pine trees and kept there needles during the winter. The majority of the trees after the turn onto my block where maple trees and had lost their leaves, combine with the large drop from the decline of the hill at this point in the winter you can see quite a long distance at the top of that hill.

As I made the turn onto my block past the trees lining the road I saw a triangular craft about 200 feet overhead I slowed down immediately and could see the object perfectly. It hovered for about 30 seconds, it was black in appearance with white lights in each of the tree corners and a red orb shaped light in the center.

The object turned after a short time to face out over the expanse that overlooked the top of the hill where I was sitting in the car (assuming the slightly longer side of the triangle was in fact the front), and began to accelerate, first slowly and then began to accelerate faster and faster until it reached incredible speed and in just a few seconds had traversed the many miles of opened visible distance until I could no longer see it.

Years later in December of 2011 I and several others witnessed the same object in the sky hovering over the main town area of Kings Park, New York again.

It was again hovering at very low altitude this time passably lower than 200 feet because the street and building lights where close enough to light up the object in the night sky, the object was huge, I was at the super market in town and saw the object after noticing several people just standing there staring at it. It stayed in the sky there for much longer this time, about a minute or more but when it left the area this time it only slowly hovered off over the tree line until it was eventually out of view.

I hope everyone who has read this book seeking answered to questions about UFO's learned a little more. All the information in this book was researched to the best of my ability.

If you are a true UFO seeker, an abductee or just someone interested to know more about UFO's then I hope you enjoyed this book, thank you for reading it and…

BE WELL…